BIOLOGY
at a Glance
Fourth edition

Judy Dodds

Illustrations by
Annette Whalley
and
Cactus Design

CRC **CRC Press**
Taylor & Francis Group
Boca Raton London New York

CRC Press is an imprint of the
Taylor & Francis Group, an **informa** business

Cover illustrations show a transverse section of the buttercup root, and a pollen grain under a scanning electron microscope. Courtesy of Bryan G. Bowes and James D. Mauseth, from *Plant Structure*, A Colour Guide, 978-1-84076-092-7, Manson Publishing.

CRC Press
Taylor & Francis Group
6000 Broken Sound Parkway NW, Suite 300
Boca Raton, FL 33487-2742

© 2014 by Taylor & Francis Group, LLC
CRC Press is an imprint of Taylor & Francis Group, an Informa business

No claim to original U.S. Government works

Printed on acid-free paper
Version Date: 20130618

Printed and bound in India by Replika Press Pvt. Ltd.

International Standard Book Number-13: 978-1-84076-199-3 (Paperback)

Visit the Taylor & Francis Web site at
http://www.taylorandfrancis.com

and the CRC Press Web site at
http://www.crcpress.com

FROM THE AUTHOR

This book offers a clear and concise approach to the teaching and learning of GCSE Biology. It covers the main biological content required by all the examining boards for both the Double Award Science and separate Biology Award, including IGCSE.

Emphasis is placed on biological principles and the application of knowledge in areas such as genetic engineering, genetic fingerprinting, fish farming and commercial uses of enzymes.

As a biology teacher for many years, I have come to realise that students learn most effectively when presented with a diagrammatic form of information. Writing notes is both tedious and non productive at all levels of ability and does not enhance understanding – a picture stays in the mind while text does not. This book aims to inform and explain by using clearly annotated diagrams, together with relevant text.

Judy Dodds

BIOLOGY IN THE NEWS

Mandy Allwood's Octuplets

After taking fertility drugs, Mandy Allwood found herself pregnant with octuplets. Going against doctors' advice, she refused to abort any of them, reportedly did a deal with the News of the World, and then miscarried all eight.

THE SECRET OF LIFE
The DNA code is finally cracked

STILL THE SLAUGHTER GOES ON
The foot and mouth outbreak continues

DNA proves remains are King Richard III

A skeleton found under a Leicester car park in 2012 has been confirmed as the remains of Richard III who died in 1485. Genetic fingerprinting matched the DNA in the skeleton to living descendants of the king. Analysis showed his skeleton had suffered 10 injuries, including 8 to the skull and his spine was badly curved. He was the last English King to die in battle.

SCIENTISTS ARE RACING TO CREATE A GENETICALLY MODIFIED 'SUPER MOSQUITO' THAT WILL DESTROY MALARIA INSTEAD OF SPREADING IT

Gene code opens new fields of medicine

Giant bat triumphs over fruit farmer...

A **north Queensland** farmer has been ordered to stop electrocuting thousands of giant bats that were feasting

Super-dingoes

Australian sheep farmers are being terrorised by packs of highly aggressive dogs, resulting from mating between dingoes and domestic dogs gone wild. The new crossbreed recently killed a young boy.

Human remains identified as the last Romanovs, by DNA testing

GENETIC SUPER BABIES STORM

Key discovery raises spectre of designer children with high IQs

White House targeted in anthrax terror campaign

First Human Embryo Cloned

Maize crop modified by GM genes

Questions:
1. Use the internet to write one page about one of the topics in the news. List your sources at the end.
   (A useful web site is www.bbc.co.uk/genes)
2. Collect articles relating to biology over the last few weeks. Stick them on a page in a similar way. Why did you choose these articles?
3. Why are people concerned about cloning?

1

CELLS

CELLS A cell is the basic unit of life. All living organisms are made of cells.

Animal and plant cells share many features but there are differences.

| Animal cell | (Shared features) | Plant cell |

Cell membrane
This holds the cell together.
It controls what enters
and leaves a cell.

Chloroplast
This contains
chlorophyll
which absorbs
light energy for
photosynthesis.

Nucleus
This controls all
cell activity. It
contains chromosomes
which control inherited features,
i.e. it carries genetic information.

Cellulose cell wall
This gives structural
support to the cell.

Cytoplasm
Here all cell activity
takes place, e.g. respiration.

Large vacuole
This contains water and
dissolved substances
together called cell sap.
This gives the cell support,
making it firm or turgid.

In addition, cells have little organelles called mitochondria which are the site of aerobic respiration, and ribosomes, where proteins are made in the process called protein synthesis.

	Animal cells	Plant cells
Features in common	Have a nucleus. Have a cell membrane. Have cytoplasm.	Have a nucleus. Have a cell membrane. Have cytoplasm.
Differences	Do not have a cell wall. Do not have chloroplasts. Do not have a large vacuole.	Have a cell wall made of cellulose. Have chloroplasts. Have a large vacuole filled with cell sap.

The size of a cell is limited by the distance over which diffusion is efficient.

Questions:
1. State two differences between animal and plant cells.
2. What is the function of the cell membrane?
3. Which three features do animal and plant cells share?
4. When plant and animal cells are placed in water, most animal cells will burst, whereas plant cells will not. Explain this difference.
5. Where does photosynthesis take place in a plant cell?

VARIETY OF CELLS

Animal cells

1. Red blood cell

A ———— B

Side view

A ———— B

This has no nucleus. It contains **haemoglobin** which combines with oxygen. Red blood cells carry oxygen round the body.

2. Cheek cell

A simple unspecialised animal cell.

3. Sperm cell

Tail

The tail helps the sperm swim towards the egg.

4. Egg (ovum)

Food can be stored in this large cell.

5. Intestine cell

Food

Folds (**microvilli**) These increase the surface area for absorption of food. The cells are found in the small intestine.

6. Tracheal cell

This **ciliated cell** is found in the trachea where the cilia hairs sweep up mucus and any bacteria and dust that enter, helping to keep the lungs clean.

Cilia hairs

7. Neurone (nerve cell)

Messages from other neurones

These long cells quickly carry messages round the body.

Direction of impulse

Muscle

Plant cells

1. Palisade cell (in leaf)

Cytoplasm

Cellulose cell wall

Cell membrane

Large vacuole (cell sap)

Chloroplasts These contain **chlorophyll**. Here **sugar** is made by **photosynthesis** during the day.

Nucleus

2. Root hair cell

This **root hair cell** has a large surface area for anchorage and absorption of water and minerals. Lots of water can enter quickly through the large surface area. There are no chloroplasts in root cells as there is no light in the soil.

Large vacuole with cell sap

Cellulose cell wall

Root hair

Cytoplasm

Cell membrane

Nucleus

Water enters

Simple organism, e.g. *Amoeba*

Nucleus

Cytoplasm

Cell membrane

This one-celled organism lives in freshwater ponds.

Questions:
1. Why do sperm cells have a tail?
2. Root hair cells in plants have a large surface area. How does this help?
3. Cells in the trachea (windpipe) have cilia hairs. What is their job?
4. Can you suggest why red blood cells are pale in the middle?
5. Why must the ovum be larger than the sperm cell?

3

LEVELS OF ORGANISATION

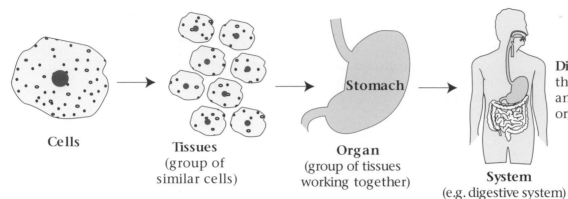

Cells

Tissues
(group of
similar cells)

Stomach

Organ
(group of tissues
working together)

System
(e.g. digestive system)

Digestive system includes the stomach, oesophagus and intestines, i.e. different organs working together.

Systems work together in an organism.

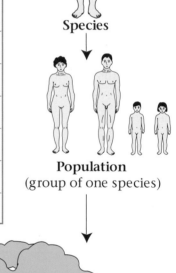

Organism
(member of
a species)

Species

Population
(group of one species)

There are nine major systems in the human body		
System		Function
Digestive		To digest and absorb food.
Breathing		To take oxygen into the body and remove carbon dioxide.
Excretory		To remove waste materials from the body.
Circulatory		To carry blood round the body.
Nervous		To carry messages round the body.
Sensory		To receive information.
Muscle		To bring about movement.
Skeletal		To provide support, protection and movement.
Reproductive		To produce young.

From Bowes: A Colour Atlas of Plant Propagation

Woodland ecosystem.

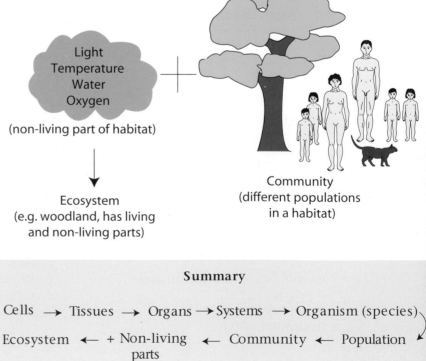

Light
Temperature
Water
Oxygen

(non-living part of habitat)

Ecosystem
(e.g. woodland, has living
and non-living parts)

Community
(different populations
in a habitat)

Summary

Cells → Tissues → Organs → Systems → Organism (species)

Ecosystem ← + Non-living ← Community ← Population
parts

HOW SUBSTANCES ENTER A CELL

- Diffusion.
- Active transport.
- Osmosis.

1. Diffusion   This is the movement of molecules from a region where they are in **high** concentration to a region where they are in **lower** concentration. Diffusion continues until the molecules are evenly mixed and there is no difference in concentration.

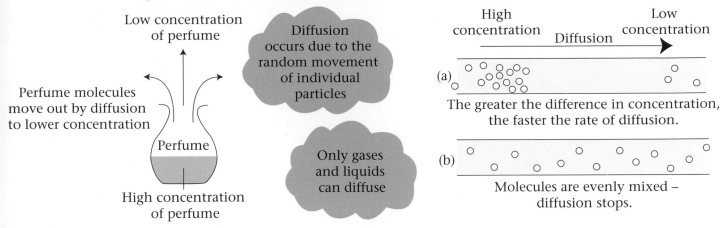

Low concentration of perfume

Diffusion occurs due to the random movement of individual particles

Perfume molecules move out by diffusion to lower concentration

Perfume

High concentration of perfume

Only gases and liquids can diffuse

High concentration   Low concentration   Diffusion

(a)

The greater the difference in concentration, the faster the rate of diffusion.

(b)

Molecules are evenly mixed – diffusion stops.

2. Active transport   Molecules move from an area of *low* concentration to an area of *high* concentration (opposite to diffusion).

Low concentration   Active transport   High concentration

This requires energy

In kidney tubules, glucose passes into blood by diffusion and active transport.

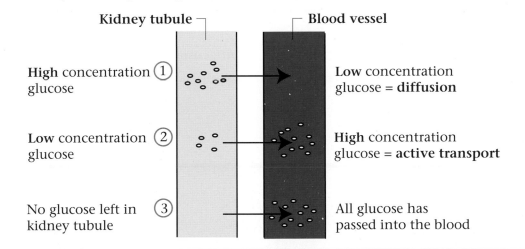

Kidney tubule   **Blood vessel**

High concentration ① glucose

Low concentration glucose = **diffusion**

Low concentration ② glucose

High concentration glucose = **active transport**

No glucose left in ③ kidney tubule

All glucose has passed into the blood

Example of active transport in plants

Root hair cells are able to absorb mineral ions from the soil by active transport.

Uptake of minerals

Minerals can enter by **diffusion** and **active transport** (low to high concentration).

Low nitrate

High nitrate

Nitrates in

Nitrates can enter the root hair cell by **active transport** which requires energy so the concentration in the plant may be higher than in the soil.

Root hair cell

Questions:
1. A drop of ink in water will spread until all the liquid is blue. What is this process called?
2. How is diffusion involved in attracting insects for pollination?

OSMOSIS The movement of water through a membrane

Osmosis is the movement of *water* from an area of **high water concentration** to an area of **lower water concentration** through a **selectively permeable membrane**.

WATER

High water concentration ——————————⟶ Lower water concentration

Water molecules are constantly moving due to kinetic energy. Solutes, like sugar, attract water molecules making them less free to move. Therefore solutes affect the ability of water to move. The more solute molecules present, the less free water molecules are to move.

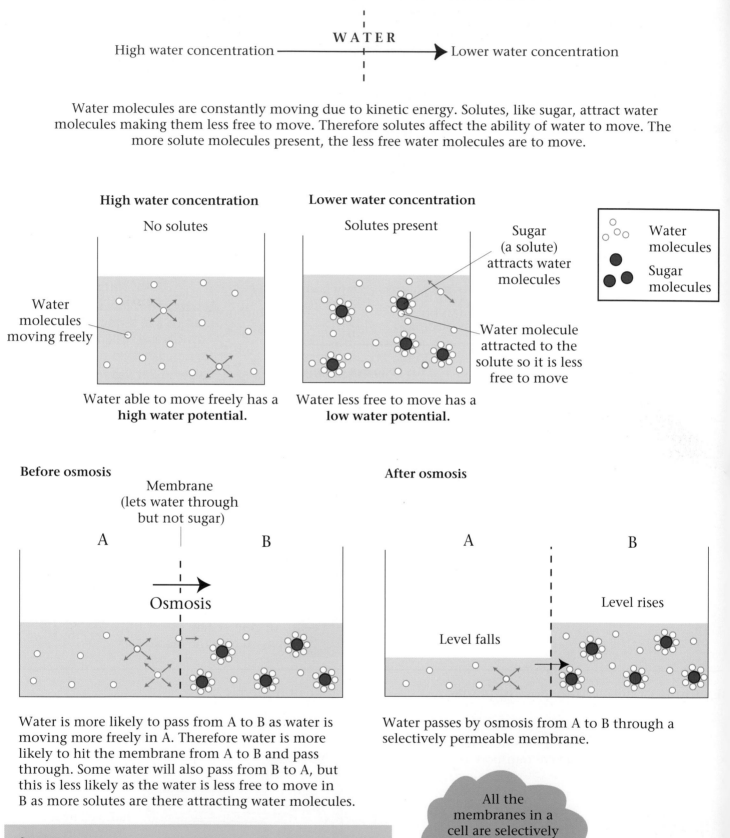

High water concentration

No solutes

Water molecules moving freely

Water able to move freely has a **high water potential.**

Lower water concentration

Solutes present

Sugar (a solute) attracts water molecules

Water molecule attracted to the solute so it is less free to move

Water less free to move has a **low water potential.**

Water molecules

Sugar molecules

Before osmosis

Membrane (lets water through but not sugar)

A   B

Osmosis

Water is more likely to pass from A to B as water is moving more freely in A. Therefore water is more likely to hit the membrane from A to B and pass through. Some water will also pass from B to A, but this is less likely as the water is less free to move in B as more solutes are there attracting water molecules.

After osmosis

A   B

Level rises

Level falls

Water passes by osmosis from A to B through a selectively permeable membrane.

All the membranes in a cell are selectively permeable.

Question:
1. A girl watered her pot plants with sea-water instead of fresh water, thus adding solutes to the soil. The plants wilted and died. Using osmosis, can you explain why?

OSMOSIS IN ACTION

1. *Amoeba*, a single-celled organism that lives in freshwater ponds.

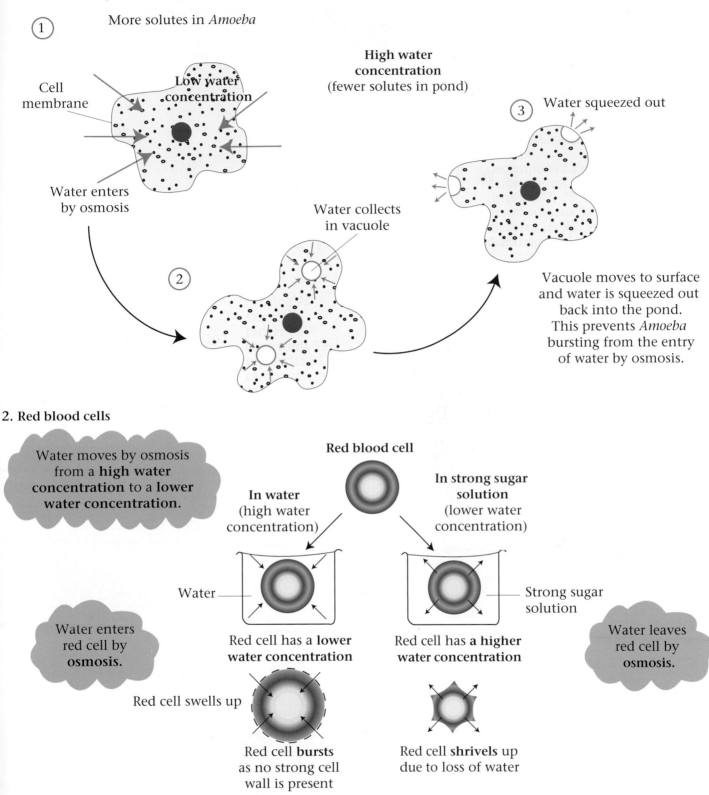

2. Red blood cells

Water moves by osmosis from a **high water concentration** to a **lower water concentration**.

Red blood cell

In water (high water concentration)

In strong sugar solution (lower water concentration)

Water

Strong sugar solution

Water enters red cell by **osmosis**.

Water leaves red cell by **osmosis**.

Red cell has a **lower water concentration**

Red cell has **a higher water concentration**

Red cell swells up

Red cell **bursts** as no strong cell wall is present

Red cell **shrivels** up due to loss of water

Questions:
1. Why must plasma (the liquid in which red blood cells are found) have the same water concentration as red and white blood cells?
2. What problems will organisms face if they live in the sea which has a lower water concentration than many organisms?

OSMOSIS AND PLANT CELLS

Plant cells do not burst when water enters by osmosis due to their strong cellulose cell wall. However, the vacuole in plant cells may lose or gain water by osmosis.

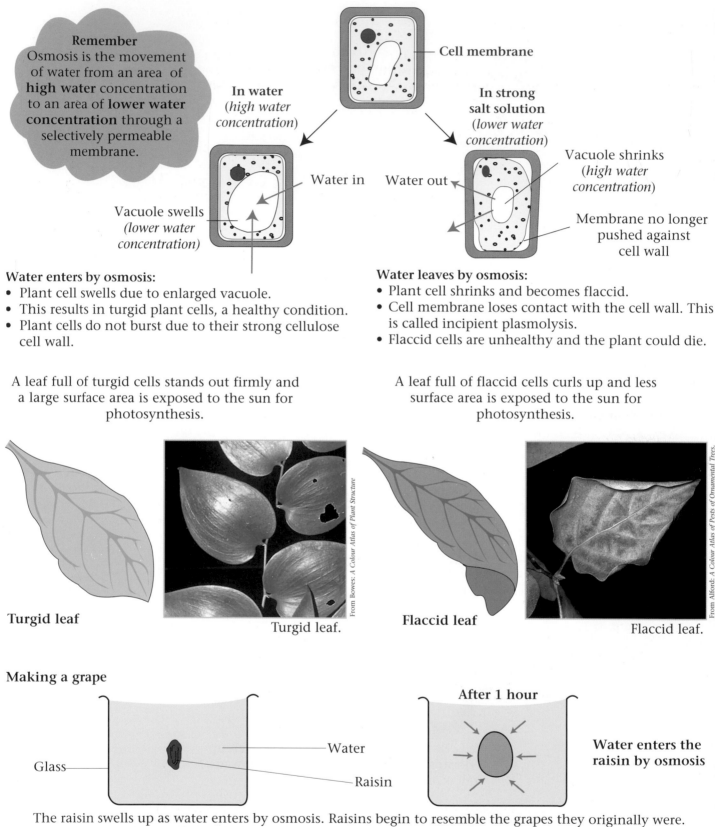

Remember
Osmosis is the movement of water from an area of **high water** concentration to an area of **lower water concentration** through a selectively permeable membrane.

Cell membrane

In water
(*high water concentration*)

In strong salt solution
(*lower water concentration*)

Water in

Vacuole swells
(*lower water concentration*)

Water out

Vacuole shrinks
(*high water concentration*)

Membrane no longer pushed against cell wall

Water enters by osmosis:
- Plant cell swells due to enlarged vacuole.
- This results in turgid plant cells, a healthy condition.
- Plant cells do not burst due to their strong cellulose cell wall.

Water leaves by osmosis:
- Plant cell shrinks and becomes flaccid.
- Cell membrane loses contact with the cell wall. This is called incipient plasmolysis.
- Flaccid cells are unhealthy and the plant could die.

A leaf full of turgid cells stands out firmly and a large surface area is exposed to the sun for photosynthesis.

A leaf full of flaccid cells curls up and less surface area is exposed to the sun for photosynthesis.

Turgid leaf

Turgid leaf.

From Bowes: *A Colour Atlas of Plant Structure*

Flaccid leaf

Flaccid leaf.

From Alford: *A Colour Atlas of Pests of Ornamental Trees, Shrubs and Flowers*

Making a grape

Glass

Water

Raisin

After 1 hour

Water enters the raisin by osmosis

The raisin swells up as water enters by osmosis. Raisins begin to resemble the grapes they originally were. Raisins do not burst due to the strong cellulose cell wall present in plants.

Questions:
1. If a piece of raw potato is placed in a strong salt solution, what do you think will happen and why?
2. Pot plants were watered with a salt solution by mistake. What do you think will happen to the plants?

THE IMPORTANCE OF VOLUME AND SURFACE AREA

Surface area is the amount of surface an organism has. If we removed our skin, flattened and measured it, this would be our surface area.

Volume is the space taken up by an organism. Large organisms take up more space, so have larger volumes.

One-celled organisms like *Amoeba,* are able to get all the oxygen they need by simple **diffusion**, i.e. oxygen moves from a higher concentration outside the cell to a lower concentration inside.

Oxygen(O_2) diffuses through the surface area of *Amoeba* and can reach into its small volume so every part of the cell gets oxygen. This is possible as *Amoeba* has a **large surface area** and a **small volume**.

Similarly, carbon dioxide (CO_2) diffuses out to the lower concentration outside the cell

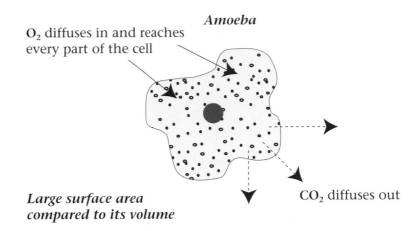

Amoeba

O_2 diffuses in and reaches every part of the cell

Large surface area compared to its volume

CO_2 diffuses out

Large organisms have two major problems with gaining oxygen:

1. Their surface area is small compared to a large volume, so insufficient oxygen enters.

The surface area for gas exchange surface is increased by the development of a folded gas exchange surface, e.g. alveoli in humans, gills in fish.

Human thorax

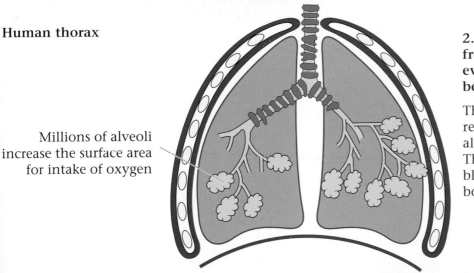

Millions of alveoli increase the surface area for intake of oxygen

2. With a large volume, the distance from the gas exchange surface to every cell is too far for diffusion to be efficient.

Therefore a **transport system**, blood, is required to carry **oxygen** efficiently to all cells and to remove **carbon dioxide**. The development of a **heart** enabled blood to be pumped all round the body.

Transport system

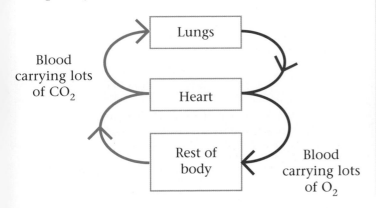

Blood carrying lots of CO_2

Lungs

Heart

Rest of body

Blood carrying lots of O_2

Questions:
1. Why is it possible for one-celled organisms to get all their oxygen by diffusion?
2. What problems do large animals face when getting oxygen and removing carbon dioxide?
3. Why must gas exchange organs be well supplied with blood vessels?
4. What other feature of gas exchange surfaces increase the uptake of oxygen?
5. Our alveoli are moist. Why, in terms of water, is it necessary for alveoli to be deep inside the body?

SURFACE AREA TO VOLUME RATIO

1 cm cube (1 cm³)

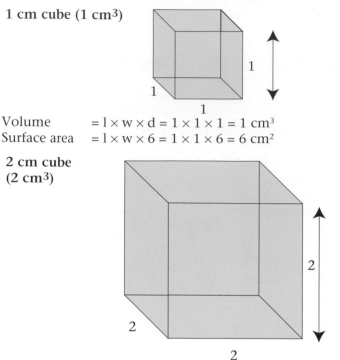

Volume = l × w × d = 1 × 1 × 1 = 1 cm³
Surface area = l × w × 6 = 1 × 1 × 6 = 6 cm²

2 cm cube (2 cm³)

Volume = l × w × d = 2 × 2 × 2 = 8 cm³
Surface area = l × w × 6 = 2 × 2 × 6 = 24 cm²

Volume – space taken up = length × width × depth
Surface area – outer surface = length × width × 6 (6 sides of a cube)

Cube size (cm)	l × w × d Volume (cm³)	l × w × 6 Surface area (cm²)	
1	1	6	Small animals have a *large* surface area compared to volume.
2	8	24	
3	27	54	
4	64	96	
5	125	150	
6	216	216	
7	343	294	Large animals have a *small* surface area compared to volume.

Questions:
1. Where is heat made in an organism and in what process?
2. How is heat lost from an animal?
3. What features of an animal will increase heat loss?
4. How would you recognise an animal living in a cold climate?
5. How would you describe the surface area to volume ratio of a) a very small animal, b) a large animal?
6. What problems do large animals face if living in hot climate and why?

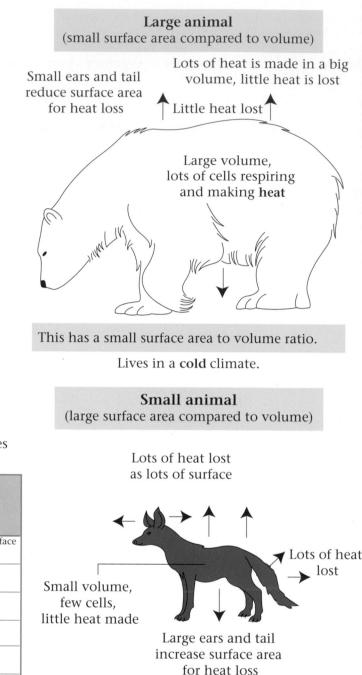

Large animal
(small surface area compared to volume)

Small ears and tail reduce surface area for heat loss

Lots of heat is made in a big volume, little heat is lost

Little heat lost

Large volume, lots of cells respiring and making **heat**

This has a small surface area to volume ratio.

Lives in a **cold** climate.

Small animal
(large surface area compared to volume)

Lots of heat lost as lots of surface

Lots of heat lost

Small volume, few cells, little heat made

Large ears and tail increase surface area for heat loss

This has a large surface area to volume ratio.

Lives in a **hot** climate.

Volume – heat is made
Heat is *made* in our cells which make up our volume, in *respiration*. Big volume = lots of heat made.

Surface area – heat is lost
Heat is *lost* through our skin or *surface area*. Big surface area = lots of heat lost.

BIOLOGICAL MOLECULES

There are three important biological molecules:
1. Proteins. 2. Lipids. 3. Carbohydrates.

PROTEINS These contain the elements carbon, hydrogen, oxygen, nitrogen.

Protein is needed for **growth**.

Sources of protein – for humans
• Meat (beef, lamb, chicken, pork).
• Fish.
• Egg-white.
• Beans.

Plants have to make their protein in order to grow. They need to combine the four elements carbon (C), hydrogen (H), oxygen (O), and nitrogen (N). Carbon, hydrogen, and oxygen are available from H_2O and CO_2. **Nitrogen** is acquired from the soil in the form of **nitrates**. **Protein** is built up from amino acids linked together by **peptide** bonds.

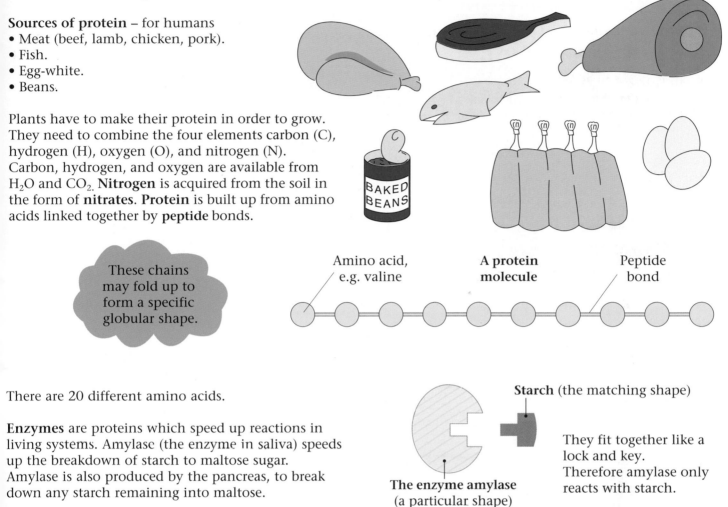

These chains may fold up to form a specific globular shape.

Amino acid, e.g. valine   **A protein molecule**   Peptide bond

There are 20 different amino acids.

Enzymes are proteins which speed up reactions in living systems. Amylase (the enzyme in saliva) speeds up the breakdown of starch to maltose sugar. Amylase is also produced by the pancreas, to break down any starch remaining into maltose.

Starch (the matching shape)

The enzyme amylase (a particular shape)

They fit together like a lock and key. Therefore amylase only reacts with starch.

Kwashiorkor

Children who do not have enough protein in their diet fail to grow properly. In parts of Africa, children may suffer from Kwashiorkor (protein deficiency). They are recognised by stick-like arms and legs and swollen abdomen, due to the build-up of tissue fluid, caused by lack of protein in their plasma.

This child exhibits thin limbs and a swollen belly, classic symptoms of kwashiorkor, severe protein deficiency. This child, although looking like an infant, is probably 3–4 years old. Although superficially looking fat, this is a form of malnutrition.

CARBOHYDRATES These contain the elements carbon, hydrogen, oxygen.

Carbohydrates include:

- An insoluble energy store (starch, glycogen).
- Soluble sugars to transport to cells for respiration.
- The cellulose cell wall in plants.

There are two main sources of carbohydrates:

Sugars
Sugar, jam, honey, treacle
(all sweet, soluble, and form crystals)

Honey

Sugar

Jam

Treacle

Starch
Bread, potatoes, rice, pasta
(all not sweet, insoluble, and do not form crystals)

Potatoes

Rice

Pasta

Bread

Carbohydrates are made of sugar units, like glucose, joined together by **glycosidic bonds.**

Sugars (one or two glucose units)

One glucose unit — Monosaccharide, e.g. glucose

Two glucose units — Disaccharide, e.g. sucrose

Glycosidic bond

Polysaccharides (many glucose units),
e.g. starch, glycogen, cellulose

Glucose units

Plants produce sugar during **photosynthesis**, combining:

$$6CO_2 + 6H_2O \longrightarrow C_6H_{12}O_6 + 6O_2$$

Carbon dioxide   Water   Glucose   Oxygen

Glucose is **respired** to release **energy** in both animals and plants.

Animal carbohydrates

- Glucose in blood.
- Glycogen in liver and muscles.

Plant carbohydrates

- Sucrose in phloem.
- Sugar in nectar.
- Starch in leaves.
- Cellulose cell wall.

Fructose is a monosaccharide used as a sweetener in the food industry. It is very sweet, so only small quantities are needed.

LIPIDS These contain the elements carbon, hydrogen, oxygen.

Lipids include fats and oils – fats are **solid** at room temperature, oils are **liquid** at room temperature.

Lipids are needed for:

- Insulation (to stop heat loss).
- Energy store.
- Waterproofing, e.g. waxy cuticle on leaves.
- Buoyancy, e.g. in large aquatic mammals.

Lipids are made up of glycerol and fatty acids joined by ester bonds.

Sources of lipids:

- Butter, margarine, vegetable oil, egg-yolk.
- Nuts are also high in lipids.
- Seeds high in lipids, e.g. sunflower seeds, corn, and soya beans, all provide us with a source of oil.

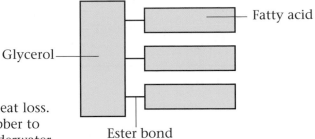

Insulation
A layer of fat under the skin acts as an **insulator** reducing heat loss. Whales and dolphins (mammals) have lots of fat called blubber to reduce heat loss in cold seas because fur cannot insulate underwater.

Waterproofing
Lipid forms the waxy cuticle covering leaves and the waxy outer layer of insects and other arthropods. In both cases, the lipid waterproofs the surface and prevents loss of water, essential for life on land.

Buoyancy
The fat stored also provides buoyancy, helping these large, heavy mammals to float in the water.

Fin whale

Sea birds are covered in a natural oil that prevents their feathers becoming waterlogged. When there is an oil spillage from a ship, detergents are often used to disperse the oil, but it also removes the natural oil from the sea birds. As a result their feathers absorb water making the birds heavy and they may drown.

From: Summerhayes and Thorpe: *Oceanography, An Illustrated Guide*, courtesy of Tony Martin.

Hunted fin whale brought to whaling station.

Saturated fats increase blood cholesterol levels. Mono-unsaturated and polyunsaturated fats may help to reduce the blood cholesterol levels.

Find out which fats are saturated and which are polyunsaturated.

Questions:
1. What are the three biological molecules?
2. Which elements are found in all three molecules?
3. Which biological molecule contains nitrogen?
4. Why is protein needed by organisms? Give a good source of protein for humans.
5. What is the function of a layer of fat under the skin?
6. What are the two main groups of carbohydrates?
7. In what form is sugar transported in a) plants, b) animals?

From: Summerhayes and Thorpe: *Oceanography, An Illustrated Guide*, courtesy of Professor Olof Linden/ICCE.

Large blowout oil spill from oil rig in the Gulf of Mexico.

13

FOOD TESTS

Test for	Chemical reagent	Colour change
Glucose	**Benedicts** solution	Blue → orange/brown
Starch	**Iodine** solution	Red → blue/black
Protein	**Biuret** reagent	Blue → purple/mauve
Lipid	**Ethanol** and water	Mixture → white

A **balanced diet** should contain a mixture of proteins, lipids, carbohydrates, vitamins, and minerals.

1. Test for glucose

Benedicts

Blue ⟶ Orange/brown

Blue Benedicts
Solution to test
Heat
Orange/brown colour

An orange/brown colour means that **glucose** is present.

2. Test for starch

Iodine

Red/brown ⟶ Blue/black

Few drops iodine
Substance to test
Blue/black colour

A blue/black colour means **starch** is present.

3. Test for protein

Biuret

Blue ⟶ Purple/mauve

Biuret solution
Solution to test
Purple/mauve colour

A purple colour means that **protein** is present.

4. Test for lipid (fats + oils)

Ethanol + Test substance

Water ⟶ White

LIQUID
Add a few drops of water
Ethanol
Butter
SHAKE MIXTURE
Water
White emulsion

Pour liquid from ① into ②

A white colour means that **lipid** is present.

Food	Benedicts	Ethanol	Biuret	Iodine
Peanuts	Blue	White	Purple	Red
Bread	Blue	Clear	Blue	Blue/black
Egg	Blue	White	Purple	Red
Apple	Orange	Clear	Blue	Blue/black
Meat	Blue	White	Purple	Red

Questions:
1. Which foods (left) contain glucose?
2. What food types does egg contain, and can you explain why?
3. Protein is found in which foods?
4. Eating which foods would give you a mixture of glucose, fat, and protein?

ENZYMES

Enzymes control the **rate** of a reaction. They are **biological catalysts**, speeding up reactions in living organisms.

Enzyme action

① Enzyme, e.g. amylase
Substrate, e.g. starch
Active site
(where the enzyme and substrate touch)

② **Enzyme–substrate complex**
They fit together like a lock and key.

③ Products, e.g. maltose
The enzyme has broken down the substrate. The enzyme can be used again.

Enzymes are **specific.** They only speed up one reaction by joining with the matching substrate.

Enzymes and temperature

Enzymes and pH

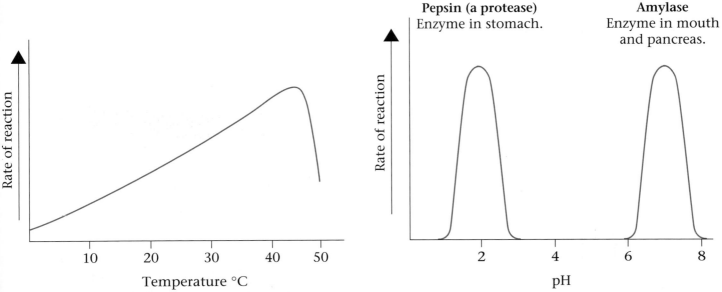

Pepsin (a protease) Enzyme in stomach.
Amylase Enzyme in mouth and pancreas.

As the temperature rises, the enzyme and substrate molecules gain kinetic energy and move faster. This increases the chance of collision between them and so increase the rate of reaction.

Most enzymes work best at 40°C. Above 40°C, their shape changes and they no longer fit with their substrate. They are denatured or destroyed. Denaturing is irreversible.

Enzymes inside living cells speed up the processes of respiration, photosynthesis and protein synthesis.

Enzymes work best at their particular pH range.

Enzymes:
- Speed up reactions – they are known as biological catalysts.
- Are specific.
- Work best at 40°C.
- Are made of protein.
- Work at a particular pH.
- Are not used up in a reaction.
- Most names end in -ase.

Questions:
1. What are enzymes made of?
2. At what temperature do enzymes work best in animals?
3. Why do enzymes stop working at high temperatures?
4. How do enzymes affect the rate of reactions?
5. Why is the shape of enzymes important?
6. What is the name of the substance to which the enzyme attaches?

Enzymes speed up the digestion of food.
There are three main types of digestive enzymes:
1. **Amylases** break down starch (into maltose).
2. **Lipases** break down lipids (into fatty acids and glycerol).
3. **Proteases** break down proteins (into amino acids).

COMMERCIAL USES OF ENZYMES

Enzymes can be mass produced in factories and are used to produce:
- **Biological washing powders.**
- **Fructose** – a sweetener in the food industry.
- **Clinistix** – to detect diabetes.

Enzymes are both **specific** and **sensitive**

Their particular shape only allows reaction with a matching shaped substrate, it is **specific** to that substrate.

Enzymes can react with tiny amounts of substrate, i.e. they detect small quantities. They are **sensitive**.

---

1. Biological washing powders

Clothes are stained by proteins (blood, meat, egg-white) and fat (oil, grease, egg-yolk). A protein-digesting enzyme, a protease, and a fat-digesting enzyme, a lipase, are needed. These enzymes are present in biological washing powders to clean our clothes effectively.

---

2. Production of fructose – a sweetener

Extracting sugar from sugar cane is expensive. To produce large quantities of sugar cheaply, enzymes are used.

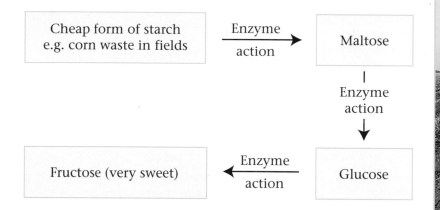

Cheap form of starch e.g. corn waste in fields	Enzyme action →	Maltose

Maltose → Enzyme action → Glucose

Fructose (very sweet) ← Enzyme action ← Glucose

As fructose is so sweet, little is required and profits are high. Fructose is used in fruit drinks, cake mixes and pie fillings.

Corn crop.

Courtesy of Holt Studios international

---

3. Detection of diabetes (caused by lack of the hormone insulin.)

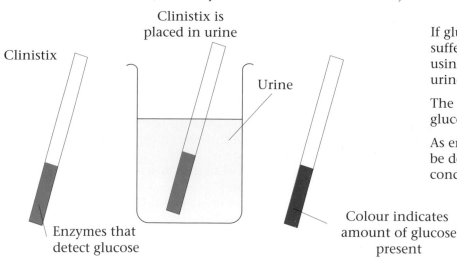

Clinistix

Clinistix is placed in urine

Urine

Enzymes that detect glucose

Colour indicates amount of glucose present

If glucose is present in the urine, a person suffers from **diabetes**. This can be detected using **Clinistix**, which is dipped into the urine.

The resulting colour indicates how much glucose, if any, is present.

As enzymes are sensitive, tiny quantities can be detected allowing early treatment of the condition.

VARIATION AND INHERITANCE

VARIATION

We belong to a species called **Homo sapiens** (humans). We do not look the same – there are many differences between us. These differences are caused in two ways.

1. Features caused by the environment (not inherited)

Cause	Effect
Sun	Sunburn
Accident	Scar
Weight lifting	Powerful muscles
Over-eating	Lots of fat
Lack of food	Poor growth

These features are caused by our way of life; they are not inherited. You cannot pass these features to your children. They are **acquired**, not inherited.

Permed hair
Dyed hair
Pierced ears
Broken nose (accident)
Scar

2. Features caused by genes (inherited)

Hair colour

Nose shape

Skin colour

Height

Freckles

Tongue rolling

Eye colour

These features are **inherited** and pass to us from our parents. We will pass these on to our children in our **genes**.

A gene is a section of DNA found on chromosomes.

Children **inherit** these features from their parents. This is why members of a **family** look **similar**.

Hair colour and type
Eye colour
Shape of nose
Freckles
Tongue rolling ability

Genes or environment?
Intelligence, sporting ability and health are determined by both genetic and environmental factors.
Which is more important is debatable.

Questions:
1. Give two features that are inherited and not affected by the environment.
2. Name two features which will be affected by the environment.
3. Name one feature that may be affected by both our genes and the environment.
4. Who do we inherit our features from?
5. What is the biological name for humans?

CAUSES OF GENETIC VARIATION

1. Formation of sex cells (meiosis)

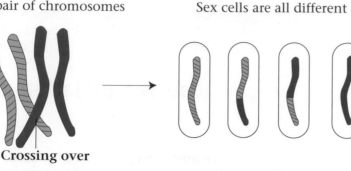

One pair of chromosomes

Crossing over

Sex cells are all different

Crossing over of chromosomes during meiosis leads to new combinations of genes. Sex cells are all genetically different. This causes **variation** between sex cells and **variety** in future offspring.

The main causes of variation

- **Meiosis** – formation of gametes, crossing over and independent assortment lead to variation.
- **Fertilisation.**
- **Mutations** lead to new features not present before.
- Meiosis and fertilisation occur during sexual reproduction. Therefore **sexual reproduction** causes variation.

2. Fertilisation

Joining of sperm and egg combines unique features.

Egg   **Sperm**   **Fertilised egg**

The fertilised egg has genes from both parents. This new mixture causes variation in the new offspring. Some will be 'fitter' than others and more likely to **survive**.

3. Mutation

Radiation causes

Mutation ← Chemical causes

Natural causes

A **mutation** is a sudden change in a gene or chromosome.

Mutations occur naturally, but the frequency of mutations can be increased by radiation or chemicals.

Usually, **mutations** are **harmful** and cause problems; occasionally, they are of great benefit to the organism.

Mutations cause changes to the DNA making up a gene, so altering the gene. Mutations can also change the number of chromosomes in a cell – both lead to genetic variation.

Mutations can cause the following harmful conditions:	
Sickle cell anaemia	
Cystic fibrosis	Once the mutation has occurred, the change is passed on to future children. It is inherited.
Haemophilia	
Huntington's disease	

The gene causing **haemophilia** appeared by mutation in Queen Victoria, affecting most of the royal families of Europe (see page 28).

Useful mutations help the organism. Mutations have caused some **bacteria** to be **resistant** to **antibiotics**, so increasing their chances of survival and reducing ours.

Questions:
1. What are the two main causes of variation?
2. Which type of variation is passed on to our children?
3. How does the formation of gametes lead to variation?
4. How does the process of fertilisation lead to more variation?
5. What is a mutation?
6. What factors cause mutations?
7. Can mutations be helpful to organisms? Give one example.

Mutations always cause variation or **change**.

DNA (DEOXYRIBONUCLEIC ACID)

DNA is found in the nucleus, making up genes. As genes determine all our features and cell activity, DNA is essential for life.

DNA, or genes, determine what proteins are made by cells. This includes enzymes which control all cell activity. As DNA can copy itself, genes can be inherited.

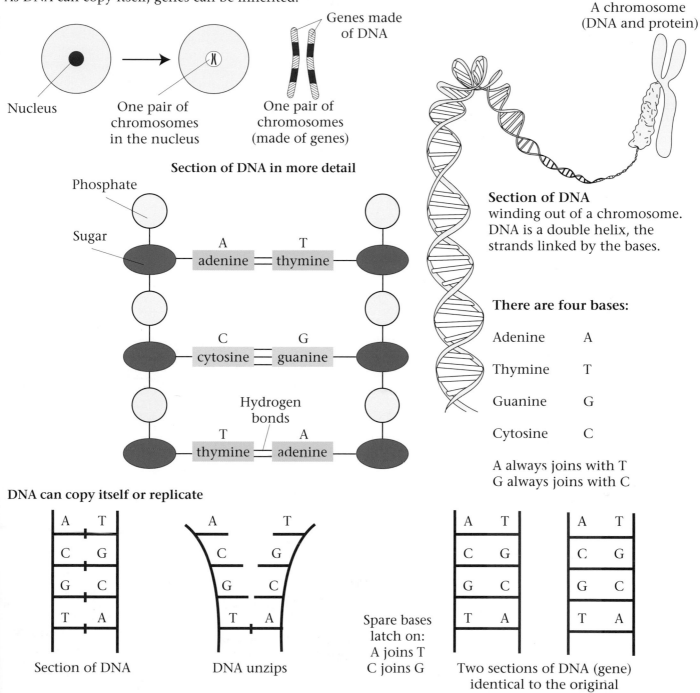

Nucleus

One pair of chromosomes in the nucleus

One pair of chromosomes (made of genes)

Genes made of DNA

A chromosome (DNA and protein)

Section of DNA in more detail

Phosphate

Sugar

A adenine ═ T thymine

C cytosine ═ G guanine

Hydrogen bonds

T thymine ═ A adenine

Section of DNA winding out of a chromosome. DNA is a double helix, the strands linked by the bases.

There are four bases:

Adenine   A

Thymine   T

Guanine   G

Cytosine   C

A always joins with T
G always joins with C

DNA can copy itself or replicate

Section of DNA

DNA unzips

Spare bases latch on:
A joins T
C joins G

Two sections of DNA (gene) identical to the original

Daughter cells produced by mitosis can therefore have identical DNA to the parent. Replication of DNA also enables genes to pass on to the sex cells, in other words to be inherited.

DNA can be extracted from cells such as onion and kiwi fruit.

Find out about the Human Genome Project.

Questions:
1. Where is DNA found?
2. What is made of DNA?
3. Why is DNA essential for life?
4. DNA controls the formation of proteins. Why is this so important?
5. What four bases make up DNA?
6. Which bases join?
7. Why is the replication of DNA so important?
8. Describe how DNA replicates.

CELLS AND CHROMOSOMES

1. Cells

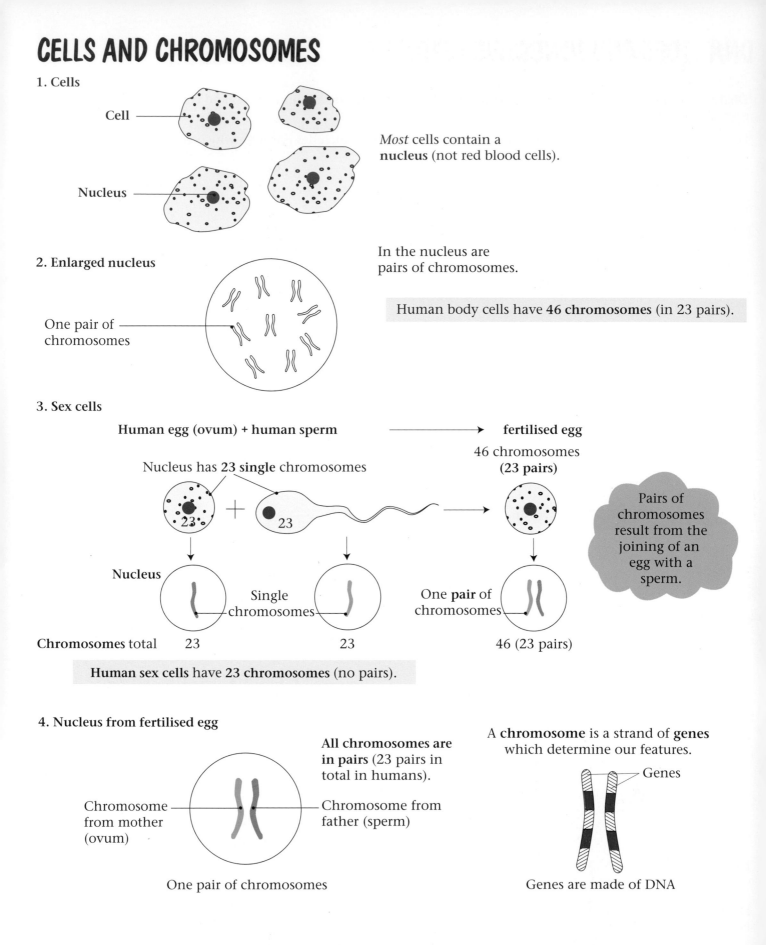

Cell

Nucleus

Most cells contain a **nucleus** (not red blood cells).

2. Enlarged nucleus

One pair of chromosomes

In the nucleus are pairs of chromosomes.

Human body cells have **46 chromosomes** (in 23 pairs).

3. Sex cells

Human egg (ovum) + human sperm ⟶ **fertilised egg**

46 chromosomes (23 pairs)

Nucleus has **23 single** chromosomes

23 + 23

Nucleus

Single chromosomes

One **pair** of chromosomes

Pairs of chromosomes result from the joining of an egg with a sperm.

Chromosomes total   23   23   46 (23 pairs)

Human sex cells have **23 chromosomes** (no pairs).

4. Nucleus from fertilised egg

All chromosomes are in pairs (23 pairs in total in humans).

Chromosome from mother (ovum)

Chromosome from father (sperm)

One pair of chromosomes

A **chromosome** is a strand of **genes** which determine our features.

Genes

Genes are made of DNA

Questions:
1. How many chromosomes are there in human body cells? How are they arranged?
2. Why do the sex cells have half the usual number of chromosomes?
3. What happens to the chromosome number at fertilisation?

DOMINANT AND RECESSIVE FEATURES

Nucleus from fertilised egg

Chromosome from mother (ovum) —— Chromosome from father (sperm)

One pair of chromosomes

Allele for brown eyes — Allele for blue eyes — This gene controls eye colour.

Blonde — Black — This gene controls hair colour.

One allele — One allele

*This will produce **brown** eyes and **black** hair as these are **dominant** features.*

Every cell, apart from the sex cells, has a full set of genes. Only some of these genes are used in any one cell.

Alleles are different forms of a **gene**. Eye colour depends on what alleles are present, e.g. brown, blue, green, grey. Normally two alleles (one on each of a pair of chromosomes) determine our features.

There are two different **alleles** here controlling eye colour, brown and blue. This person will have brown eyes, as brown is **dominant** (stronger) to blue. We call blue **recessive**, or weaker. Two alleles determine our features, one from each parent. The stronger, dominant, allele will show.

B represents black hair
b represents blonde hair

This person will have black hair, it is dominant.

Dominant alleles are shown by a capital letter, e.g. T.
Recessive alleles are shown by a lower case letter, e.g. t.

Dominant features	Recessive features
Black hair	Blonde hair
Curly hair	Straight hair
Brown eyes	Blue eyes
No freckles	Freckles

Courtesy of Dick Williams

Courtesy of D. Colgan

Dominant and recessive alleles have an equal chance of being inherited.

Questions:
1. What does dominant mean?
2. What is an allele?
3. What alleles would produce blue eyes?
4. Why are chromosomes found in pairs?
5. How many alleles usually determine a feature?
6. What does recessive mean?

GENETIC CROSSES (I)

1. Black hair × blonde hair

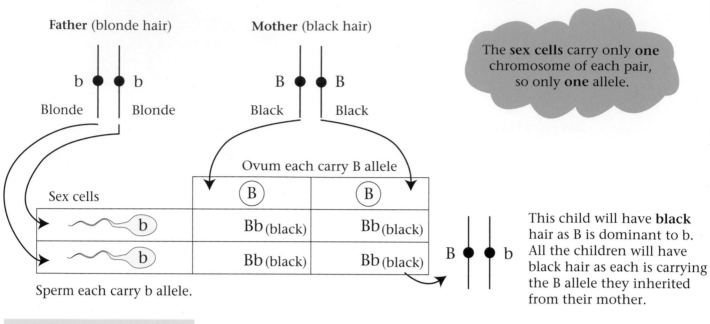

Father (blonde hair)   **Mother** (black hair)

b — Blonde   b — Blonde   B — Black   B — Black

The **sex cells** carry only **one** chromosome of each pair, so only **one** allele.

Ovum each carry B allele

Sex cells

	B	B
b (sperm)	Bb (black)	Bb (black)
b (sperm)	Bb (black)	Bb (black)

Sperm each carry b allele.

This child will have **black** hair as B is dominant to b. All the children will have black hair as each is carrying the B allele they inherited from their mother.

B ● ● b

2. Black hair × black hair

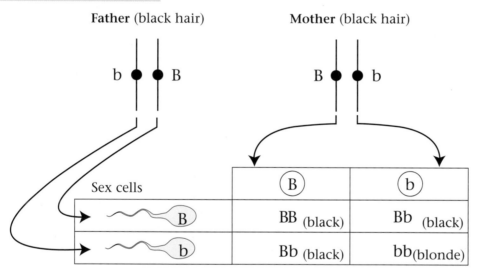

Father (black hair)   **Mother** (black hair)

b ● ● B   B ● ● b

Sex cells

	B	b
B (sperm)	BB (black)	Bb (black)
b (sperm)	Bb (black)	bb (blonde)

When the sex cells join in fertilisation, pairs of chromosomes are formed and pairs of alleles. The **dominant** allele determines the colour.

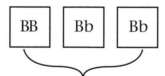

| BB | Bb | Bb | | bb |

All black hair, as B is dominant   Blonde hair

This gives a three black to one blonde ratio or **3:1**, i.e. there is a 1 in 4 chance of these parents having a blonde child

---

Genetic terms

Chromosome	A strand of genes.
Gene	A section of DNA controlling a feature.
Allele	A different form of a gene.
Dominant	Stronger allele (capital letter).
Recessive	Weaker allele (small letter).
Genotype	The type of alleles present.
Phenotype	The appearance of the organism.
Homozygote	Two alleles are the same.
Heterozygote	Two alleles of a pair are different.

Questions:

1. What type of letters are used for dominant features?
2. What colour hair would the following genotypes produce (B = black hair, b = blonde hair): BB, Bb, bb?
3. What colour hair would the children of two blonde parents have and why?

GENETIC CROSSES (II)

Tongue rolling is caused by a **dominant allele** R . The recessive allele r = non-tongue roller. 84% of the population can tongue roll.

There are three genotypes possible:

RR	Rr	rr
Tongue roll	Tongue roll	Cannot tongue roll

84%   16%

Problem
Two parents, both of whom can tongue roll have a child who cannot. What is the genotype of the parents?

> **Gregor Mendel** was a monk working in the garden of the monastery in Brno, now in the Czech Republic. He studied the inheritance of features in pea plants. From his studies the basic laws of inheritance were discovered in 1865. However, his ideas were not appreciated until the early 1900's as there was little understanding of genes and chromosomes at that time.

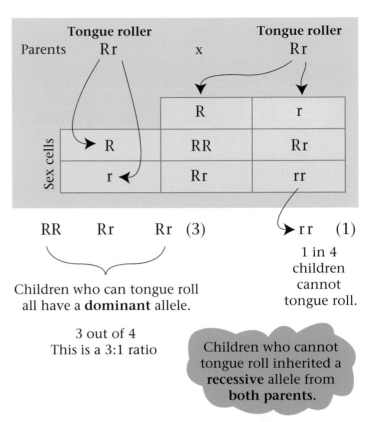

RR   Rr   Rr   (3)   ➤ r r   (1)

Children who can tongue roll all have a **dominant** allele.

3 out of 4
This is a 3:1 ratio

1 in 4 children cannot tongue roll.

> Children who cannot tongue roll inherited a **recessive** allele from **both parents**.

Pea plants
Height is controlled by a pair of alleles: T (Tall) and t (short). Tall pea plants are TT or Tt. Short pea plants are tt.

Two alleles are the **same** (homozygote).

1. **Tall**   ×   **Short**

2. **Tall**   ×   **Short**

Two alleles are **different** (heterozygote).

The pea plants were either tall or short, as found in the original parents. No pea plants of intermediate height were produced and therefore no 'blending' of alleles occurred.

Questions:
1. Freckles are caused by a recessive allele. Would the following people have freckles: FF, Ff, ff?
2. What is a homozygote? Give an example.
3. How is it possible to find out if a pea plant is a homozygote or heterozygote?
4. If both parent pea plants are tall but have a short offspring, what must be the genotype of the parents?

CYSTIC FIBROSIS A *recessive* inherited disorder.

Three possible genotypes:

FF	Ff	ff
Healthy	Healthy but a carrier	Cystic fibrosis sufferer (caused only by the double recessive)

Effects of cystic fibrosis

Thick mucus blocks air tubes in lungs:
- Stops air reaching lungs.
- Allows infection as bacteria breed on mucus.
- Causes coughing.

Thick mucus blocks pancreatic duct:
- Stops digestive enzymes leaving the pancreas.
- Food in duodenum is not digested properly.
- Little food is absorbed.

Gene therapy is being used in an attempt to reduce the symptoms of cystic fibrosis. However, targeting the specific cells is difficult and side-effects may be a problem.

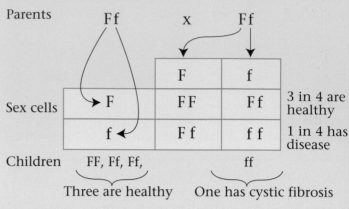

Two healthy parents, if carriers, have a 1 in 4 or 25% chance of having a child with cystic fibrosis.

Parents: Ff x Ff

Sex cells	F	f
F	FF	Ff
f	Ff	ff

3 in 4 are healthy
1 in 4 has disease

Children: FF, Ff, Ff, — Three are healthy | ff — One has cystic fibrosis

1 in 20 people in Europe are carriers and perfectly healthy.
1 in 2,000 babies is born with cystic fibrosis which may lead to an early death.

HUNTINGTON'S DISEASE A *dominant* inherited disorder.

Three possible genotypes:

HH	**Hh**	**hh**
Very rare, probably lethal	Huntington's disease (lethal later)	Healthy

As it is caused by a *dominant* gene, only one dominant allele is needed to cause the condition. The condition develops over the age of 40 years.

	Sufferer parent (no symptoms yet)	Healthy parent
Parents	Hh x	hh

Sex cells	h	h
H	Hh	Hh
h	hh	hh

2 in 4 has disease
2 in 4 are healthy

Children: Hh, Hh — Two will develop Huntington's disease | hh, hh — Two will be healthy

Half the children are likely to inherit the harmful 'H' gene

Effects of Huntington's disease (in the over-40s)
Causes jerky, erratic movements. Mental powers reduce very quickly. Sufferers die quickly once the disease starts. Unfortunately, they may already have passed the condition to their children. Genetic counselling for families with conditions such as Huntington's disease may be advisable to explain the risks of having affected children.

Polydactyly (having more than 5 fingers) is another dominant, inherited disorder transmitted in the same way as Huntington's disease.

Polydactyly.

SICKLE CELL ANAEMIA A *recessive* inherited disorder.

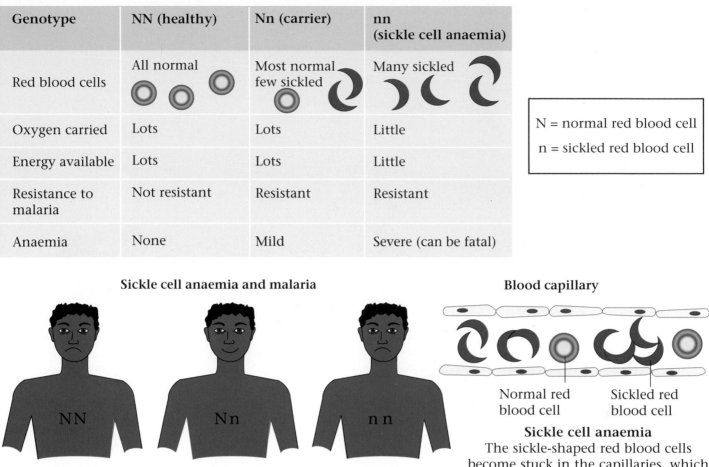

Genotype	NN (healthy)	Nn (carrier)	nn (sickle cell anaemia)
Red blood cells	All normal	Most normal few sickled	Many sickled
Oxygen carried	Lots	Lots	Little
Energy available	Lots	Lots	Little
Resistance to malaria	Not resistant	Resistant	Resistant
Anaemia	None	Mild	Severe (can be fatal)

N = normal red blood cell

n = sickled red blood cell

Sickle cell anaemia and malaria

Blood capillary

Normal red blood cell   Sickled red blood cell

NN
Killed by malaria
(healthy in non-malarial regions)

Nn
Resistant to malaria
(mild anaemia)

n n
Killed by severe anaemia

Sickle cell anaemia
The sickle-shaped red blood cells become stuck in the capillaries, which causes painful, swollen joints and deprives body cells of oxygen. Sufferers may die at an early age.

People with the genotype **Nn** are common in malarial regions of the world as this gives **resistance to malaria**. Unfortunately, this increases the number that suffer from sickle cell anaemia, nn.

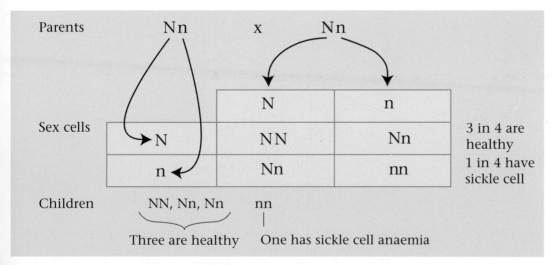

If both the mother and father are **carriers**, there is a 1 in 4 chance of their having a child with **sickle cell anaemia**.

Parents   Nn   x   Nn

	N	n
N	NN	Nn
n	Nn	nn

Sex cells

3 in 4 are healthy
1 in 4 have sickle cell

Children   NN, Nn, Nn   nn

Three are healthy   One has sickle cell anaemia

Questions:
1. What shape can red blood cells be?
2. Draw the shape of red blood cells found in,
   a) healthy people?
   b) people suffering from sickle cell anaemia?
3. How do sickle-shaped red blood cells affect individuals?
4. Describe the symptoms of sickle cell anaemia.
5. What is meant by a carrier for sickle cell anaemia?
6. Why is it an advantage for people to be a carrier for sickle cell anaemia in certain parts of the world?
7. If both parents are carriers, what % of their children may suffer from sickle cell anaemia? Explain this in a genetic diagram.
8. Why do you think there are so many people with sickle cell anaemia in Uganda, Tanzania and Kenya?

SEX CHROMOSOMES

Humans have 23 pairs of chromosomes, of which **one pair** determines whether we are **male** or **female**.

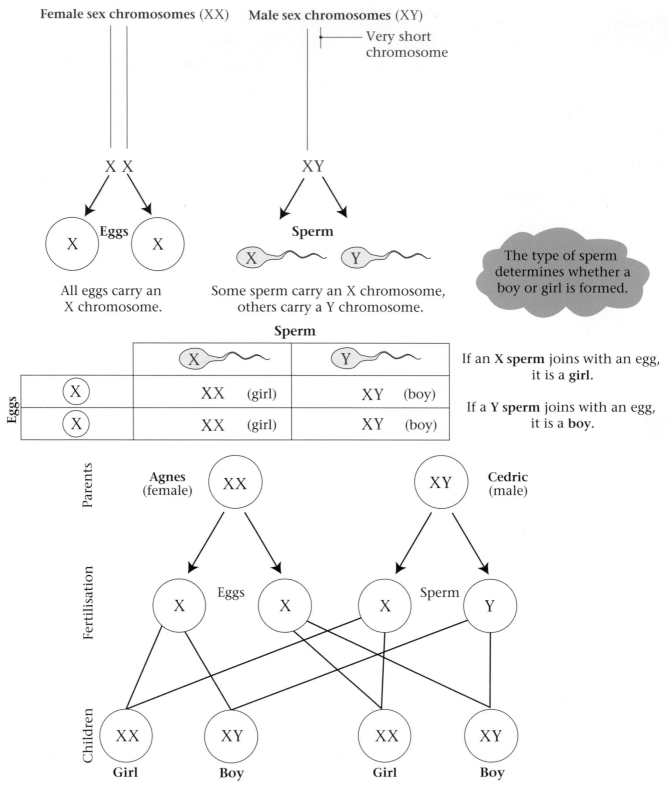

Female sex chromosomes (XX)   Male sex chromosomes (XY)

— Very short chromosome

X X   XY

X Eggs X   Sperm X   Y

All eggs carry an X chromosome.   Some sperm carry an X chromosome, others carry a Y chromosome.

The type of sperm determines whether a boy or girl is formed.

Sperm

Eggs		X	Y
	X	XX (girl)	XY (boy)
	X	XX (girl)	XY (boy)

If an **X sperm** joins with an egg, it is a **girl**.

If a **Y sperm** joins with an egg, it is a **boy**.

Parents: Agnes (female) XX   XY Cedric (male)

Fertilisation: X Eggs X   X Sperm Y

Children: XX   XY   XX   XY
          Girl   Boy   Girl   Boy

Two girls and two boys (equal), 1:1 ratio.   As there is an **equal number of X and Y sperm**, the number of girls and **boys** born is almost the same.

Each baby has a 50% chance of being male and 50% chance of being female.

Questions:

1. Which two chromosomes produce a female?
2. What are the two types of sperm?
3. What sex chromosome is always found in the egg?
4. What are the chances of having a baby boy, compared to a baby girl? Explain.
5. If a family already has five girls, what are the chances that the next baby will be a boy?
6. What determines the sex of the child?

SEX-LINKED CHARACTERISTICS

These are caused by **genes** usually on the **X sex chromosome**, e.g. haemophilia, colour-blindness and baldness. They are recessive genetic disorders.

Sex chromosomes

H = healthy	E = normal vision	B = not bald
h = haemophilia	e = colour blind	b = bald
(a blood disorder)		

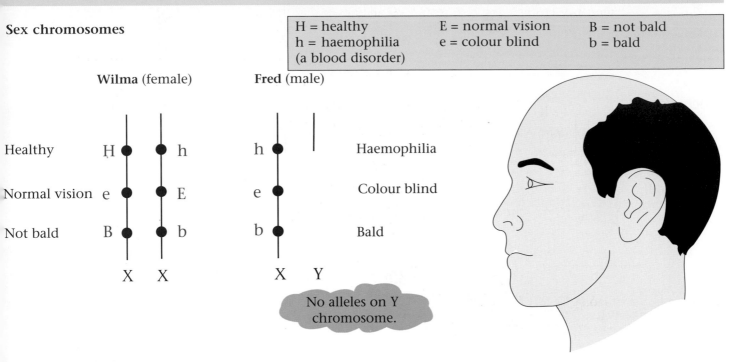

Wilma (female)

Fred (male)

Healthy   H   h   h   Haemophilia

Normal vision   e   E   e   Colour blind

Not bald   B   b   b   Bald

X   X   X   Y

No alleles on Y chromosome.

Wilma is healthy, has normal vision and will not go bald as she has dominant alleles present.

Fred has haemophilia, is colour blind and will go bald as only recessive alleles are present on the X chromosome.

Inheritance of haemophilia from Queen Victoria

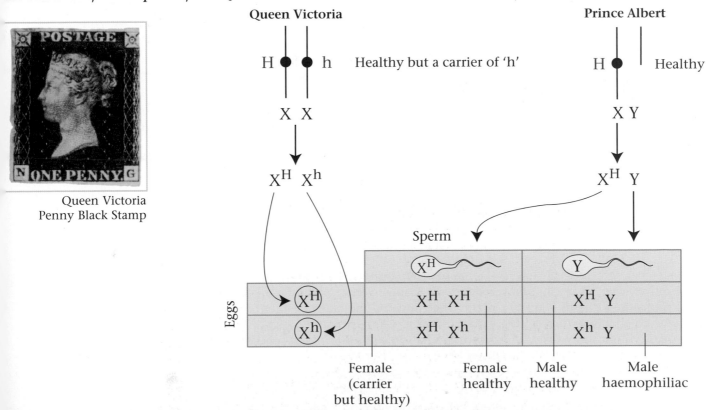

Queen Victoria Penny Black Stamp

Queen Victoria

H   h   Healthy but a carrier of 'h'

X X

$X^H X^h$

Prince Albert

H   Healthy

X Y

X^H Y

Sperm

Eggs

	X^H	Y
X^H	$X^H X^H$	X^H Y
X^h	$X^H X^h$	X^h Y

Female (carrier but healthy)   Female healthy   Male healthy   Male haemophiliac

Many descendants of **Queen Victoria** suffered from **haemophilia** and died young.
The present **British Royal Family** do **not** carry the harmful recessive gene for haemophilia.

Men are more likely to suffer from sex-linked conditions as they have no alleles on the Y chromosome to prevent recessive alleles on the X being expressed.

HAEMOPHILIA

Queen Victoria and the inheritance of haemophilia in the royal families of Europe.

Haemophilia is an example of a recessive, sex-linked, genetic disorder carried on the X sex chromosome. It leads to bleeding due to a faulty clotting mechanism. The main problem is internal bleeding into joints and muscles, which causes severe pain and damage to joints. Treatment involves replacing the missing clotting factor using blood transfusions. Unfortunately, some contaminated blood has been used and this has led to widespread infection of HIV (leading to AIDS), and Hepatitis C amongst some sufferers of haemophilia. Haemophilia is normally carried by females and passed to their sons.

Questions:
1. Why is the British royal family free of haemophilia?
2. How many people suffered from haemophilia in this family tree in total?
   a) How many were males?
   b) How many were females?
3. Can you explain the difference in the number of males and females affected?
4. Why is it not possible for males to be carriers of haemophilia?

CODOMINANCE When two alleles of a pair are equally dominant.

Parents   **Red flowers**   ×   **White flowers**
            R R   W W

RR = Red
WW = White
RW = Pink

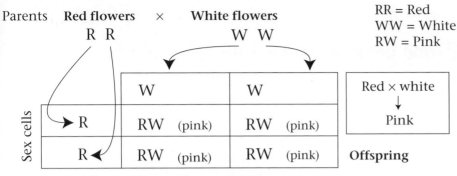

Sex cells		W	W
	R	RW (pink)	RW (pink)
	R	RW (pink)	RW (pink)

Red × white
↓
Pink

Offspring

All are RW, and have **pink** flowers.

> Both red and white
> are dominant.

If red alleles were **dominant**, flowers would be **red**.

If white alleles were **dominant**, flowers would be **white**.

A **third colour** indicates **both** are dominant = **codominance**.

Parents   **Pink flowers**   ×   **Pink flowers**
            R W   R W

Snapdragons

Sex cells		R	W
	R	RR (red)	RW (pink)
	W	WR (pink)	WW (white)

Pink × pink
↓
Red, pink, white
1:2:1

Offspring

RR   WR, RW   WW
One red   Two pink   One white

1:2:1 ratio

Blood groups

There are four blood groups:
A, B, AB, and O.
• A and B are dominant.
• O is recessive.
(Note that there are three
types of alleles: A, B, and O.)

Blood group	Genotype	% in UK
A	AA or AO	40
B	BB or BO	12
AB	AB	3
O	OO	45

Group AB is an example of **codominance**.

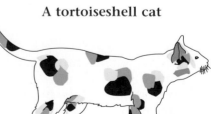

A tortoiseshell cat

Coat colour in cats is sex-linked and involves codominance.
It is carried by alleles on the 'X' sex chromosome.

As female cats are XX, they can have 2 alleles controlling coat
colour, whereas males only have one X chromosome, so only
one allele controlling coat colour.

Only females can have both a 'B' and 'Y' allele resulting in
tortoiseshell colour.

All tortoiseshell cats are female

APPLICATIONS OF GENETICS

GENETIC ENGINEERING Altering genes to make a useful product.

For example, making insulin (needed for diabetics).

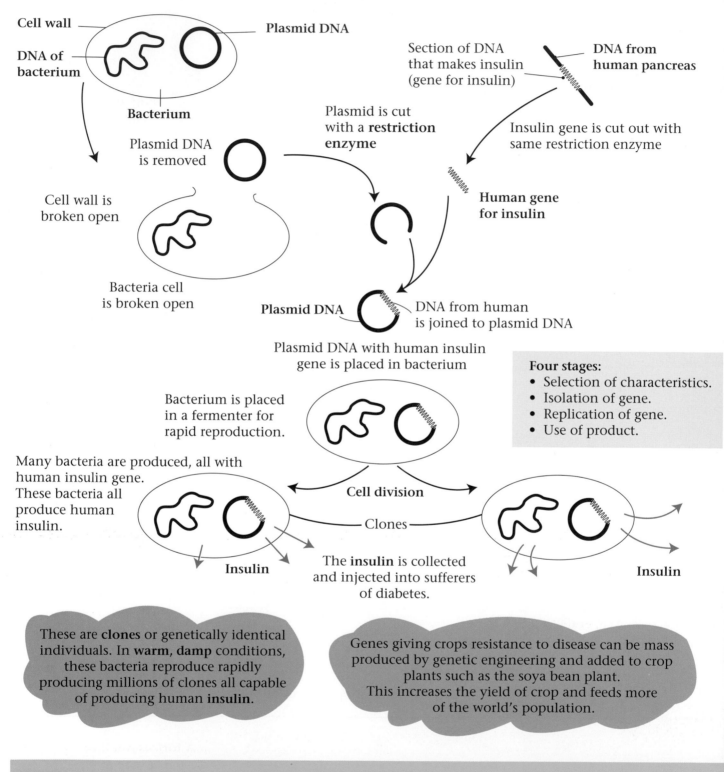

Cell wall

DNA of bacterium

Plasmid DNA

Bacterium

Plasmid DNA is removed

Cell wall is broken open

Bacteria cell is broken open

Plasmid is cut with a **restriction enzyme**

Section of DNA that makes insulin (gene for insulin)

DNA from human pancreas

Insulin gene is cut out with same restriction enzyme

Human gene for insulin

Plasmid DNA

DNA from human is joined to plasmid DNA

Plasmid DNA with human insulin gene is placed in bacterium

Four stages:
• Selection of characteristics.
• Isolation of gene.
• Replication of gene.
• Use of product.

Bacterium is placed in a fermenter for rapid reproduction.

Many bacteria are produced, all with human insulin gene. These bacteria all produce human insulin.

Cell division

Clones

Insulin

The **insulin** is collected and injected into sufferers of diabetes.

Insulin

These are **clones** or genetically identical individuals. In **warm, damp** conditions, these bacteria reproduce rapidly producing millions of clones all capable of producing human **insulin**.

Genes giving crops resistance to disease can be mass produced by genetic engineering and added to crop plants such as the soya bean plant. This increases the yield of crop and feeds more of the world's population.

Questions:
1. Which micro-organisms can be used in genetic engineering?
2. What product can be made by genetic engineering?
3. Where is the gene found that makes insulin in humans?
4. How is this human insulin gene cut out?
5. Which part of the bacterium is joined to the human gene?
6. Where are the bacteria with the human gene put to encourage rapid reproduction?

GENETIC FINGERPRINTING

Every individual has a unique pattern of DNA, (except for identical twins), inherited from both parents. This pattern can be seen as a sequence of black lines similar to a 'bar code'.

In the same way that a person's fingerprint is unique, a genetic fingerprint can be used to identify an individual. A genetic fingerprint can be used in forensic work to catch criminals from DNA left at the crime scene. It can also be used to establish paternity, where the true father is uncertain and to establish evolutionary relationships between organisms.

The pattern of DNA is **a genetic fingerprint.**

A genetic fingerprint

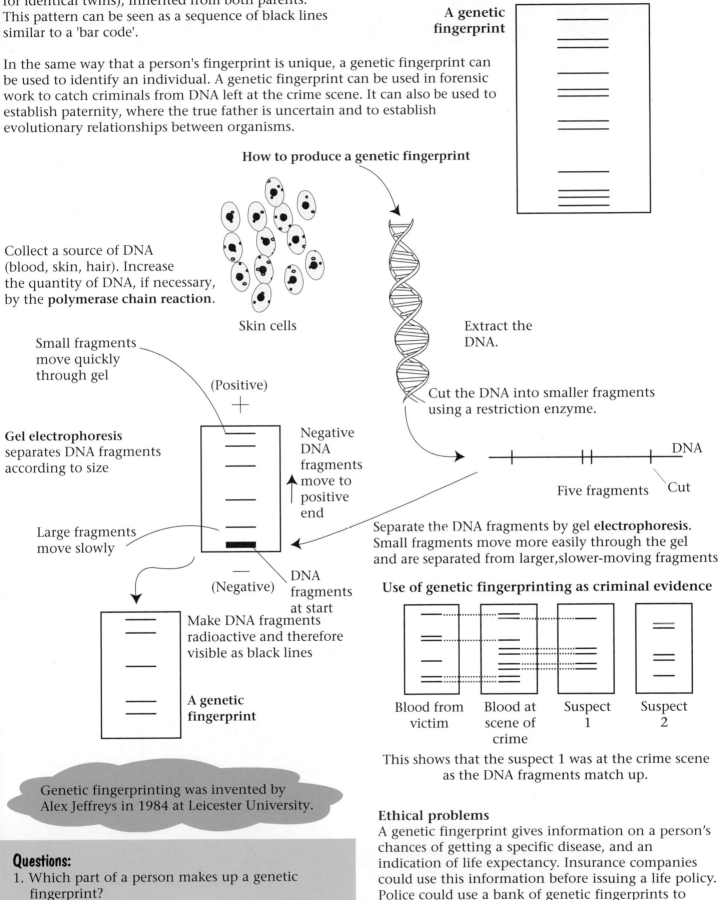

How to produce a genetic fingerprint

Collect a source of DNA (blood, skin, hair). Increase the quantity of DNA, if necessary, by the **polymerase chain reaction**.

Skin cells

Extract the DNA.

Cut the DNA into smaller fragments using a restriction enzyme.

Small fragments move quickly through gel

(Positive)
+

Gel electrophoresis separates DNA fragments according to size

Negative DNA fragments move to positive end

DNA

Five fragments   Cut

Large fragments move slowly

(Negative) — DNA fragments at start

Make DNA fragments radioactive and therefore visible as black lines

A genetic fingerprint

Separate the DNA fragments by gel **electrophoresis**. Small fragments move more easily through the gel and are separated from larger, slower-moving fragments

Use of genetic fingerprinting as criminal evidence

Blood from victim | Blood at scene of crime | Suspect 1 | Suspect 2

This shows that the suspect 1 was at the crime scene as the DNA fragments match up.

Genetic fingerprinting was invented by Alex Jeffreys in 1984 at Leicester University.

Ethical problems
A genetic fingerprint gives information on a person's chances of getting a specific disease, and an indication of life expectancy. Insurance companies could use this information before issuing a life policy. Police could use a bank of genetic fingerprints to identify criminals quickly.

Questions:
1. Which part of a person makes up a genetic fingerprint?
2. What must be extracted from cells to produce the genetic fingerprint?
3. Draw a flow diagram to show the main stages involved in producing a genetic fingerprint.

Do you think that everyone's genetic fingerprints should be taken at birth and be kept on file?

AN INDUSTRIAL FERMENTER

Fermenters provide the optimum conditions for the growth of microorganisms such as bacteria. They are like a factory for the reproduction of microorganisms.

Fermenters are large vessels made of stainless steel. Before use they are steam cleaned under pressure to kill all microorganisms. This creates sterile or ASEPTIC conditions. Fermenters are used to grow microorganisms that make a desired product. One example is growing genetically modified bacteria that produce insulin. Conditions are controlled for the maximum growth and reproduction of the specific microorganism. They need the right temperature, pH, oxygen and a supply of food. The product is collected and the fermenter is steam cleaned before being used again.

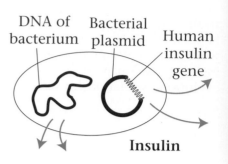

DNA of bacterium   Bacterial plasmid   Human insulin gene

Insulin

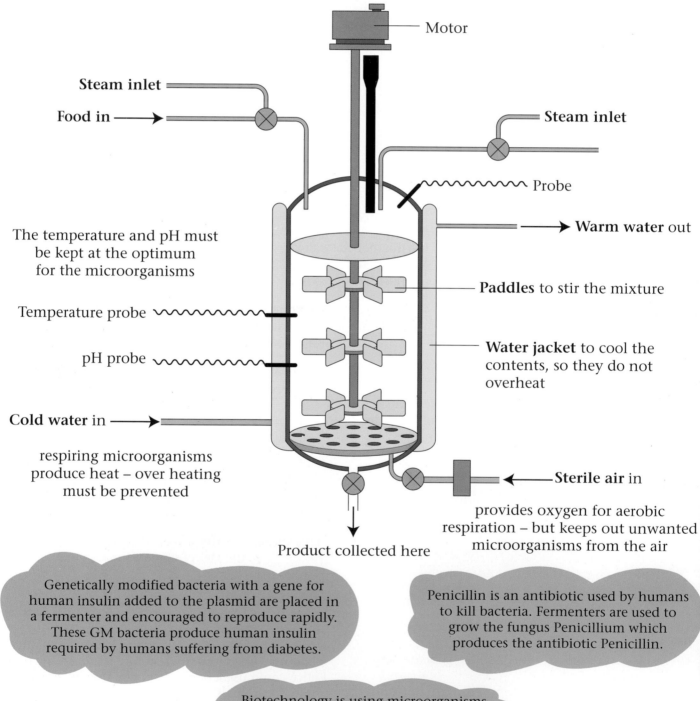

Motor

Steam inlet

Food in

Steam inlet

Probe

Warm water out

The temperature and pH must be kept at the optimum for the microorganisms

Temperature probe

pH probe

Paddles to stir the mixture

Water jacket to cool the contents, so they do not overheat

Cold water in

respiring microorganisms produce heat – over heating must be prevented

Sterile air in

provides oxygen for aerobic respiration – but keeps out unwanted microorganisms from the air

Product collected here

Genetically modified bacteria with a gene for human insulin added to the plasmid are placed in a fermenter and encouraged to reproduce rapidly. These GM bacteria produce human insulin required by humans suffering from diabetes.

Penicillin is an antibiotic used by humans to kill bacteria. Fermenters are used to grow the fungus Penicillium which produces the antibiotic Penicillin.

Biotechnology is using microorganisms to make products useful to humans e.g. Insulin made by GM bacteria

Feature of fermenter	Function
Sterile air	air provides oxygen for aerobic respiration, and being sterile prevents the entry of unwanted microorganisms, so stopping contamination.
Source of nitrogen	a source of nitrogen for microorganisms to make protein for growth e.g. amino acids, nitrates
Carbohydrate	a source of energy e.g. glucose
Cooling water jacket	to prevent overheating of the culture as respiration produces heat and lots of microorganisms respiring means more heat generated.
Temperature probe	to monitor the temperature of the culture so it can be maintained at the optimum for the microorganisms. If mixture gets too hot, enzymes denature and bacteria die. If too cold, growth and reproduction are too slow.
pH probe	to monitor the pH of the fermenter contents and keep it at the optimum for enzymes so maximising the growth of microorganisms.
Paddles	to stir the mixture, so all microorganisms get food and oxygen, and they do not sink to the bottom of the container
Steam cleaning	to create ASEPTIC (sterile) conditions i.e. the removal of all microorganisms to stop contamination. Only the required microorganisms must be present in the fermenter. Unwanted ones will contaminate the product and use up food and oxygen.

Six stages
- Steam clean the fermenter, creating aseptic conditions
- Add required microorganism
- Add source of nitrogen and energy source
- Provide oxygen in sterile air
- Monitor temperature and pH to maintain optimum conditions
- Collect required product

Measuring blood glucose level. People who suffer from diabetes have a high blood glucose level and may need regular injections of insulin to lower the glucose level back to normal. The insulin is made by genetically modified bacteria grown in an industrial fermenter.

Questions:
1. Why must the fermenter be supplied with sterile air?
2. The fermenter has temperature and pH probes, why are these necessary?
3. The inside of the fermenter is sterilised before use, why are aseptic conditions necessary?
4. Why do microorganisms need a source of nitrogen?
5. What would happen if the cooling jacket failed?
6. One bacterium can divide into two bacteria every 20 minutes in ideal conditions, as in a fermenter. Starting with one bacterium, how many bacteria will there be after 7 hours?

SELECTIVE BREEDING (artificial selection)

Selective breeding is used by animal and plant breeders to produce a **desired variety or species**. Humans select which organisms should mate.

1. In animals

This is used to produce good guide dogs for the blind.

Good female guide dog   Good male guide dog   Successful guide dog

This combines the desirable characteristics of both parents, e.g. intelligence, obedience and gentleness.

Problems with selective breeding

Selected animals and plants are similar; all of them have the same features and therefore similar genes. Breeds which are not selected for mating will die out. This leads to a **loss of genes** which might be required in the future. If a new disease emerges, it could kill all the selected animals, since all of them are similar and none may have the gene for resistance to the disease. The **loss of variation** reduces the long-term survival of a species. They are unable to adapt to changes, as the genes needed may have been lost.

2. In plants

This is used to produce a better crop and more profit for the farmers.

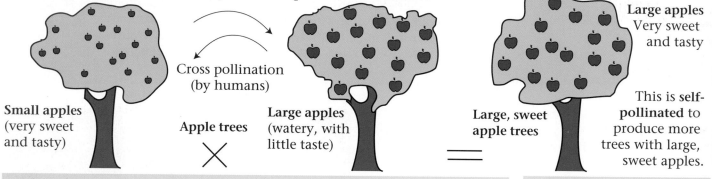

Cross pollination (by humans)

Apple trees

Small apples (very sweet and tasty)

Large apples (watery, with little taste)

Large, sweet apple trees

Large apples Very sweet and tasty

This is **self-pollinated** to produce more trees with large, sweet apples.

Pollen is transferred from one tree to the other to try to combine the good features of both, i.e. humans select which trees will 'mate' – **artificial selection**.

Any trees with small, watery apples will be **destroyed** and **not** used for **selection**. A new variety of apple tree has been produced.

Selective breeding – the process

- Choose an animal (or plant) with the desired feature, e.g. long legs, resistance to disease or a high milk yield.
- Mate with another desirable organism.
- Identify which offspring have inherited the desired gene.
- Continue to breed with *selected* offspring; do not allow the rejects to mate.
- This increases the number of animals or plants with the desired feature.

The process:
- Select feature.
- Crossbreeding
- Selection of suitable offspring over many generations.

Questions:

1. Why is selective breeding carried out?
2. When breeding cattle, what features might be selected?
   Which offspring would be used for further breeding and why?
   Which offspring would not be selected for further mating and why?
3. How is selective breeding of value to farmers?

CELL DIVISION AND EVOLUTION

CELL DIVISION Mitosis and meiosis.

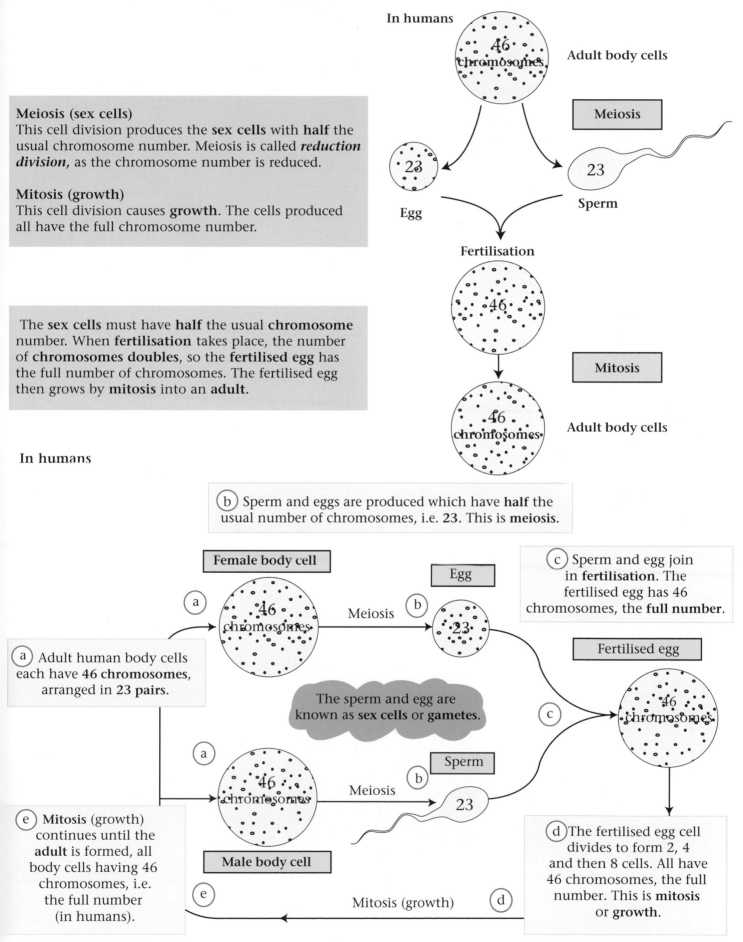

Meiosis (sex cells)
This cell division produces the **sex cells** with **half** the usual chromosome number. Meiosis is called *reduction division*, as the chromosome number is reduced.

Mitosis (growth)
This cell division causes **growth**. The cells produced all have the full chromosome number.

 The **sex cells** must have **half** the usual **chromosome** number. When **fertilisation** takes place, the number of **chromosomes doubles**, so the **fertilised egg** has the full number of chromosomes. The fertilised egg then grows by **mitosis** into an **adult**.

In humans

In humans

Adult body cells

Meiosis

Egg

Sperm

Fertilisation

Mitosis

Adult body cells

b) Sperm and eggs are produced which have **half** the usual number of chromosomes, i.e. **23**. This is **meiosis**.

Female body cell

Egg

c) Sperm and egg join in **fertilisation**. The fertilised egg has 46 chromosomes, the **full number**.

Meiosis

a) Adult human body cells each have **46 chromosomes**, arranged in **23 pairs**.

The sperm and egg are known as **sex cells** or **gametes**.

Fertilised egg

Sperm

Meiosis

e) **Mitosis** (growth) continues until the **adult** is formed, all body cells having 46 chromosomes, i.e. the full number (in humans).

Male body cell

Mitosis (growth)

d) The fertilised egg cell divides to form 2, 4 and then 8 cells. All have 46 chromosomes, the full number. This is **mitosis** or **growth**.

CELL DIVISION AND THE HUMAN LIFE CYCLE

Female reproductive system

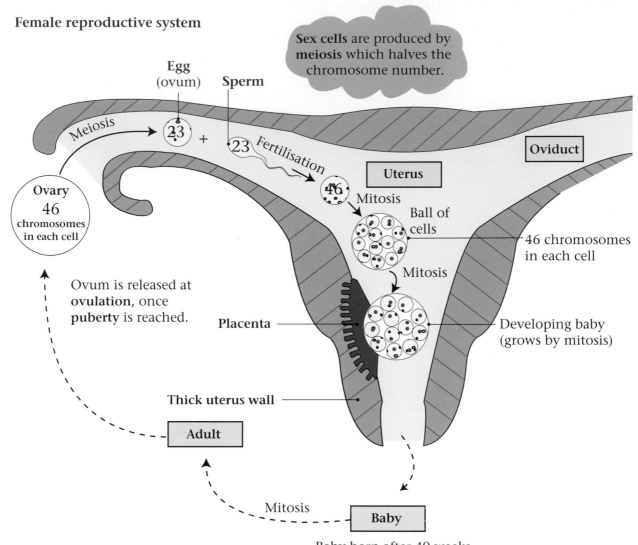

Sex cells are produced by **meiosis** which halves the chromosome number.

Egg (ovum)

Sperm

Meiosis

23 + 23 Fertilisation

Oviduct

Uterus

46 Mitosis

Ball of cells

Ovary
46 chromosomes in each cell

46 chromosomes in each cell

Ovum is released at **ovulation**, once **puberty** is reached.

Mitosis

Placenta

Developing baby (grows by mitosis)

Thick uterus wall

Adult

Mitosis

Baby

Baby born after 40 weeks

The developing baby is called an embryo. Embryo cells gradually become specialised for a particular function. Early cells in the embryo are called *embryonic stem cells*. Theses cells have the ability to become many different types of cells. Their future is not 'fixed'. Scientists are able to use embryonic stem cells to replace damaged tissue, but this raises concerns regarding embryo tissue being used for 'spare-parts'

Questions:
1. Which type of cell division produces the sex cells?
2. Why must the sex cells have half the chromosome number of the normal body cells?
3. The human fertilised egg contains how many chromosomes? Where have these chromosomes come from?
4. The fertilised egg, one cell, grows into an adult with millions of cells. Which type of cell division is this?
5. In what process does an ovary release an egg?
6. Where do the egg and sperm join in fertilisation?
7. Why is mitosis necessary in a fertilised egg?
8. How many chromosomes are in each cell produced by mitosis in humans?

MITOSIS AND MEIOSIS

Mitosis (growth)

Meiosis (reduction division forming sex cells)

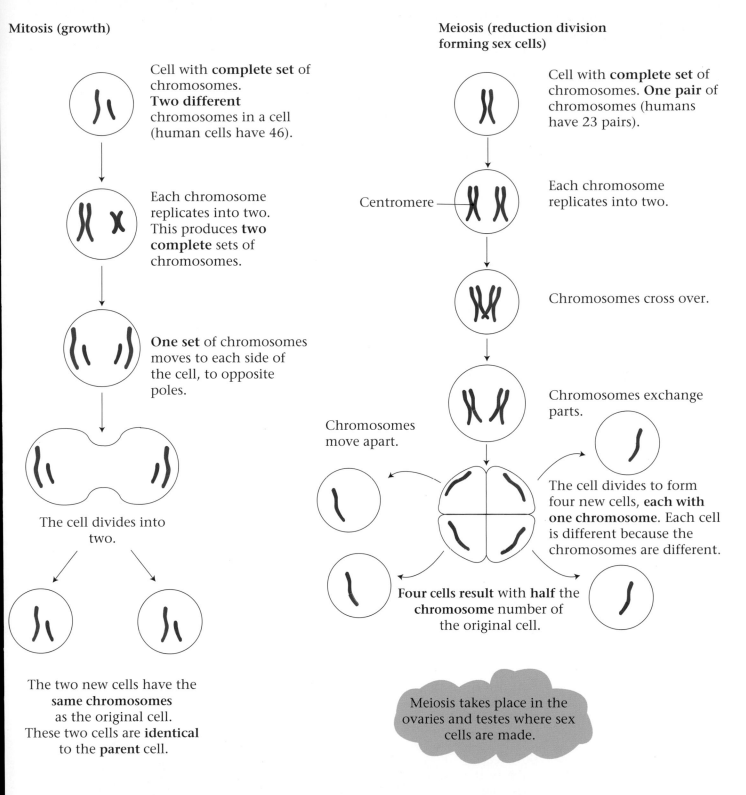

Cell with **complete set** of chromosomes. **Two different** chromosomes in a cell (human cells have 46).

Each chromosome replicates into two. This produces **two complete** sets of chromosomes.

One set of chromosomes moves to each side of the cell, to opposite poles.

The cell divides into two.

The two new cells have the **same chromosomes** as the original cell. These two cells are **identical** to the **parent** cell.

Cell with **complete set** of chromosomes. **One pair** of chromosomes (humans have 23 pairs).

Centromere

Each chromosome replicates into two.

Chromosomes cross over.

Chromosomes exchange parts.

Chromosomes move apart.

The cell divides to form four new cells, **each with one chromosome**. Each cell is different because the chromosomes are different.

Four cells result with **half** the **chromosome** number of the original cell.

Meiosis takes place in the ovaries and testes where sex cells are made.

Questions:
1. Name two places where meiosis occurs in humans.
2. How many sex cells are produced in meiosis from the parent cell?
3. What is the chromosome number in the sex cells compared with the parent cell?
4. What causes variation in the sex cells?
5. Name one place where mitosis occurs in humans.
6. How do the daughter cells in mitosis compare to the parent cell?
7. How many cells result from mitosis?

GROWTH

Growth is the permanent increase in size of an organism.

Three stages of growth

1. Cell division

2. Cell expansion (mainly in plants)
   Vacuole

3. (Animal cells)
   Nerve cell (neurone)
   Microvilli
   Cell with microvilli
   Cell specialisation

SUMMARY

Three stages of growth:
• Cell division (mitosis).
• Cell expansion.
• Cell specialisation (differentiation).

Types of growth:
• Continuous.
• Discontinuous.

Growth in most animals and plants

Height / Time

Growth slows down at old age
Rapid growth
Slow growth at first

Continuous growth

Growth in arthropods, e.g. locust

Weight / Time

Moulting

Discontinuous growth
When the outer skin is shed, insects take in water or air to expand their bodies, before new skin hardens

Human growth

Boys grow until about 18 years

Height / Age in years

Girls grow rapidly at puberty
Girls stop growing at about 16 years
Boys reach puberty at a later age

4 8 12 16 20

Cancer

When cell division is out of control

A malignant **tumour** (ball of cells) can split causing secondary tumours

Tumour (cancer)

Secondary tumours

This spreads cancer round the body

Did you know that most animals reach a particular maximum size and then stop growing, whereas plants can continue growing?

Questions:
1. What causes cell expansion in plants?
2. What kind of cell division causes growth?
3. Name four specialised cell in animals.
4. What are the three stages of growth?
5. When are girls taller than boys (refer to the human growth chart)?
6. At what age do boys grow most rapidly?

ASEXUAL REPRODUCTION IN PLANTS (NATURAL) A rapid method of reproduction, but all offspring are genetically identical to the parent (clones).

Natural clones in potato plants

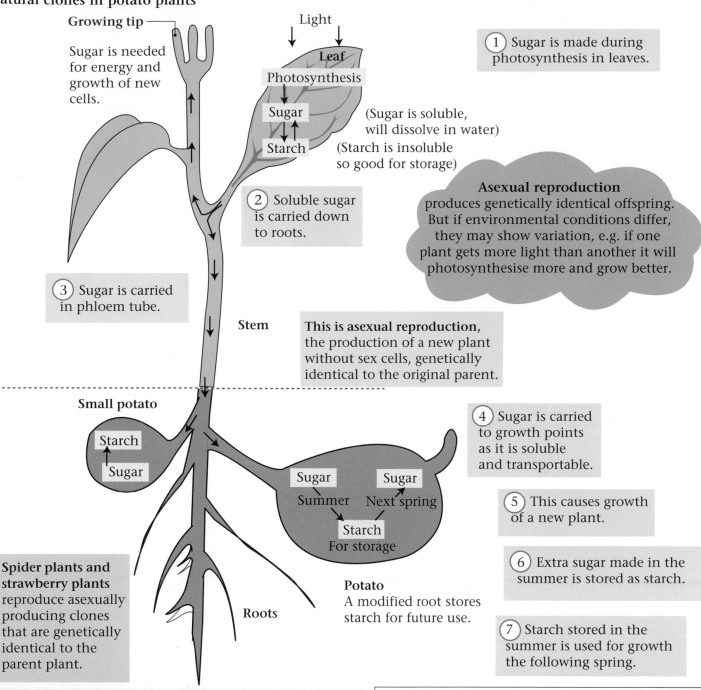

Growing tip

Sugar is needed for energy and growth of new cells.

Light

Leaf
Photosynthesis

Sugar

Starch

(Sugar is soluble, will dissolve in water)
(Starch is insoluble so good for storage)

① Sugar is made during photosynthesis in leaves.

② Soluble sugar is carried down to roots.

Asexual reproduction produces genetically identical offspring. But if environmental conditions differ, they may show variation, e.g. if one plant gets more light than another it will photosynthesise more and grow better.

③ Sugar is carried in phloem tube.

Stem

This is asexual reproduction, the production of a new plant without sex cells, genetically identical to the original parent.

Small potato

Starch

Sugar

④ Sugar is carried to growth points as it is soluble and transportable.

Sugar
Summer

Sugar
Next spring

Starch
For storage

⑤ This causes growth of a new plant.

⑥ Extra sugar made in the summer is stored as starch.

Potato
A modified root stores starch for future use.

Spider plants and strawberry plants reproduce asexually producing clones that are genetically identical to the parent plant.

Roots

⑦ Starch stored in the summer is used for growth the following spring.

Questions:
1. Asexual reproduction is reproduction without the production of gametes. As a result, there is no variation between the parent and offspring.
   Why is this, a) a good method of reproduction, b) a poor method of reproduction; compared to sexual reproduction?
2. Where is food made in plants and at what time of year?
3. How is the food transported to a storage area and in what form?
4. How can this food be used for the growth of new plants and when?
5. What evidence for the growth of new plants can be seen in a potato?

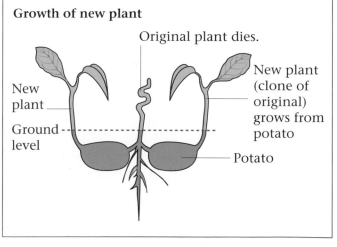

Growth of new plant

Original plant dies.

New plant

Ground level

New plant (clone of original) grows from potato

Potato

ASEXUAL REPRODUCTION IN PLANTS (ARTIFICIAL)

The production of new plants from existing ones (clones). There is no variation.

1. Cuttings from geranium

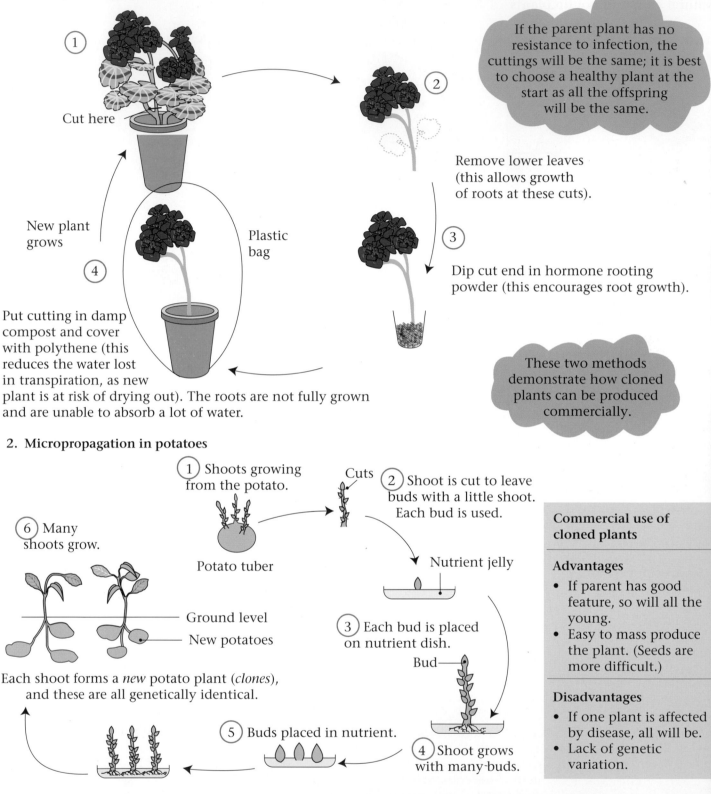

If the parent plant has no resistance to infection, the cuttings will be the same; it is best to choose a healthy plant at the start as all the offspring will be the same.

Cut here

New plant grows

Plastic bag

Remove lower leaves (this allows growth of roots at these cuts).

Dip cut end in hormone rooting powder (this encourages root growth).

Put cutting in damp compost and cover with polythene (this reduces the water lost in transpiration, as new plant is at risk of drying out). The roots are not fully grown and are unable to absorb a lot of water.

These two methods demonstrate how cloned plants can be produced commercially.

2. Micropropagation in potatoes

1. Shoots growing from the potato.

Cuts

2. Shoot is cut to leave buds with a little shoot. Each bud is used.

Potato tuber

Nutrient jelly

3. Each bud is placed on nutrient dish.

Bud

4. Shoot grows with many buds.

5. Buds placed in nutrient.

6. Many shoots grow.

Ground level

New potatoes

Each shoot forms a *new* potato plant (*clones*), and these are all genetically identical.

Commercial use of cloned plants

Advantages
- If parent has good feature, so will all the young.
- Easy to mass produce the plant. (Seeds are more difficult.)

Disadvantages
- If one plant is affected by disease, all will be.
- Lack of genetic variation.

Questions:
1. Draw a flow chart to show how new geranium plants can be produced from one plant using cuttings.
2. Why are the lower leaves removed?
3. How does dipping the cut stem in rooting powder help the process?
4. Why is the new plant covered with a plastic bag?
5. What is a clone?
6. Micropropagation involves using small pieces of the plant, such as buds. How can this be used to produce clones?
7. What is the advantage of both cuttings and micropropagation to plant growers?

40

CLONING BY TISSUE CULTURE

Summary
- Select parent plant.
- Cut off small pieces.
- Sterilise, i.e. ensure aseptic conditions.
- Growth.

Large colourful flowers

Resistant to disease

Desirable plant
with required characteristics and healthy

Weak bleach solution (sterilising solution)

Explant

Explant

Jelly with nutrients

Sterile scalpel

Stem

Explants

Each explant grows into a small plant

Enormous numbers of identical plants can be produced from small pieces of the selected parent plant.

- A healthy plant with all the desired characteristics is selected.

- The stem, or root, is cut into many small pieces, called **explants.**

- The small pieces are placed in weak bleach to sterilise them, i.e. to remove all microbes.

- The sterilised explants are placed in jelly containing all the substances required for growth, i.e. sugar, vitamins, minerals, and plant growth hormones. They are kept in a room where the light and temperature are controlled to encourage growth.

- The plants grow into clones of the parent, all genetically identical.

- The sale of these desirable plants provides a good income for the growers.

From Bowes: *A colour Atlas of Plant Propagation and Conservation.*

Explant in jelly with full range of nutrients.

41

PLANT HORMONES – AUXINS

Plant growth is controlled by hormones called **auxins**.

Experiment 1

1.

Young shoot (coleoptile) → It bends towards the light.

2. **Auxin** is concentrated here.

Auxin moves from tip and down shady side.

Auxin causes cell elongation here.

Less auxin here means **less** cell elongation.

Bending is caused by unequal distribution of auxin. More auxin causes more growth on one side of the tip.

Experiment 2

If tip cut off:

1.

No bending.
This shows that it is the *tip* that is *sensitive* to light and the *source* of the hormone.

If tip is replaced with agar jelly:

2.

Jelly

Cell elongation here.

Movement of auxin

The auxin produced in the tip must **diffuse** through the jelly and concentrate on the shady side. Here *cell elongation* causes **bending**. This shows that the **hormone** can move by **diffusion**. Light, gravity, and water cause growth movements in plants called **tropisms**. Tropisms are controlled by auxins.

Growth in young seedlings

Shoot grows up

Roots grow down

Normal growth

If placed on its side.

Shoot grows up to light (away from gravity).

Root grows away from light (towards gravity).

This is controlled by auxins (plant hormones).

Auxin causes cell elongation in both the root and shoot tips, resulting in growth there.

All plant roots bend towards the ground. They are **positively geotropic.**

Stimulus	Shoot	Root	Response called
Light	Grows towards	Grows away from	Phototropism
Gravity	Grows away from	Grows towards	Geotropism

Phototropism is a growth movement in response to light. **Geotropism** is a growth movement in response to gravity.

Auxins are used in **agriculture** to improve production of fruit. They can cause **fruit formation** in grapes without fertilisation. This results in seedless grapes. Auxins are used to encourage **root growth** in cuttings (rooting powder) and to prevent the growth of side branches from the stem. They are also used as **weed killers**. Some auxins are absorbed through leaves, so broad leaved plants absorb more causing their death. Lawns are kept free of broad leaved plants by this method. The auxins cause plants to grow rapidly, disrupting their normal growth pattern. These plants develop long weak stems and die.

Questions:
1. What is a tropism?
2. Which part of a shoot is sensitive to light? How do you know?
3. Where does the auxin concentrate in a shoot exposed to light?
4. What effect does the auxin have on the shoot?
5. What is phototropism?
6. How does phototropism help plant growth?

EVOLUTION   The changes that take place in living organisms over a long period of time.

The theory of evolution by natural selection was first proposed by Charles Darwin in 1859. It explains why animals and plants seem to fit in with their surroundings. Modern genetics confirms this theory, e.g. **Giraffes**.

1. There are too many offspring.
2. There is competition for food, space, and other things.
3. Within a species there is variation, due to differing genes.
4. Some organisms have features which help them live long enough to mate and have more young like themselves.
5. Those less fit, do not survive to breed, so their unhelpful features are not passed on.
6. The survivors are different from their ancestors because they only have the successful genes. This slight change is **evolution**.

Originally giraffes' neck length varied

With plenty of leaves to eat, all giraffes survived to breed.

Variety of neck lengths

In harsh times, leaves were left only at top of trees. Only those with long necks survived to mate with other long-necked survivors.

Only tall giraffes can feed

Originally fish colouring varied

With plenty of food and no predators all the fish types survive to breed and pass on their colouring.

Predator

Predators appear and start to eat the visible fish. Those that blend in survive better. They are camouflaged.

Weeds camouflage striped fish

Only the striped fish survive to breed and pass on the striped gene. Now all these fish are striped. The striped fish is the most fit and will survive in the weed.

Survival of the fittest

All giraffes now have long necks

The gene for a long neck is passed on to the young. The gene for shorter necks disappears with the giraffes. Eventually only giraffes with long necks remain. This is **natural selection** or **survival of the fittest** (the long-necked giraffe is the most fit and survives).

Other ideas on evolution
In 1809 Jean Baptiste Lamarck proposed that organisms acquire new features which can then be passed to their young. e.g. if giraffes constantly stretch their necks to reach leaves their longer neck will pass to their offspring. A knowledge of genetics completely disproves this idea. (Stretching does not alter genes.)

Questions:
1. What do rabbits compete for in the wild?
2. How might a longer neck help a giraffe to survive?
3. Why might only long neck giraffes remain to mate with?
4. What will their young be like and why?
5. How is this the survival of the fittest? Explain.
6. Why are the survivors different from their ancestors?
7. What happened to the gene for short necks in giraffes?

EVIDENCE FOR EVOLUTION

1. Observed natural selection

Warfarin-resistance in rats

Warfarin is used both as a rat poison and to 'thin' the blood in human's at risk from strokes.

As a rat poison, warfarin causes rats that eat it to bleed excessively and die. Warfarin is widely used and originally was very successful at reducing rat numbers. However, the few rats with resistance survived to reproduce and pass on their alleles for resistance to their young, so giving them resistance too. Rats with no resistance died out leaving only resistant ones to reproduce. Now populations of warfarin-resistant rats mean that warfarin is of no use as a rat poison. The change in the rat population over time is an example of evolution in action and has been well documented.

The resistant rats were the 'fittest', if warfarin was used. This is an example of survival of the fittest.

Rat

White peppered moth,
Biston betularia

Natural selection – the process	• Variety within one species. • Struggle for existence. • Survival of the fittest. • Most successful type emerges.

The white peppered moth is well camouflaged on pale lichen-covered tree trunks found in rural areas. It is more likely to survive to reproduce than black moths, which would be highly visible to the birds that feed on them. This leads to survival of the fittest, the light moth is camouflaged and therefore 'fittest' in country areas. The opposite applies in city areas.

2. From fossils

Camels – These appeared first in North America. Fossils show the movement and changes in camels.

First camels were in North America

X

Original camels

Modern **camels** today

Llamas today

Disease as a selection pressure
If a new disease appears, then only animals resistant to it will live long enough to breed and pass on the genes causing resistance. Non resistant animals will die out.

Disease causes selection. Only resistant animals will remain.

Original camels → Modern camels
Original camels → Llamas
This change is called **evolution**. These are now separate species unable to interbreed

Camels arose in **North America**. Some then migrated over to **Africa** and **Asia**. Other camels moved down to **South America**. Once the camels had been **isolated**, different features were **selected** for survival and the two camel populations became **different**. This movement is confirmed by the **fossil record** which is almost complete and can be dated.

Questions:
1. Why is there competition for survival?
2. Which peppered moths survive best in country areas and why?
3. What has caused the increase in the number of black peppered moths in Manchester?
4. Which two species have evolved from original camels and what is the evidence for this?
5. How can disease cause selection in animals? Which animals are more likely to survive?

SPECIES

All living organisms have two names (genus and species)

	Genus	Species	
e.g.	*Homo*	*sapiens*	(humans)
	Urtica	*dioica*	(stinging nettles)
	Lumbricus	*terrestris*	(earthworms)

Definition
Members of a species look alike and can interbreed successfully to produce fertile offspring.

Only members of the **same** species can breed and produce **fertile** offspring.

Parents (same species)   *Homo sapiens*

Mating possible and successful.

Children

Fertile offspring

As mating is successful and produces fertile offspring they are the same species, *Homo sapiens*.

Parents (different species)

Horse   Donkey

Mating possible but this produces mules.

Offspring   Mule

Infertile offspring unable to breed.

Mules are infertile so a horse and donkey are not the same species.

In the fossil record
Other similar species to humans are found:

(Genus)	(species)
Homo	*erectus*
man	upright

Found in Africa and Asia less than a million years ago, they shared the same genus, *Homo*, as humans, but not the same species.

How fossils may forms

- Hard parts of an organism may be preserved such as bones and teeth.
- A mould of an organism may be embedded in a rock.
- The entire body may be preserved, e.g. insects in resin, but this is rare.

The cat family share many features:

Same **genus**	All different **species**	
Panthera	*leo*	lion
Panthera	*tigris*	tiger
Panthera	*pardus*	leopard

Parents

Tiger   x   Lion   **Mating may be possible.**

No fertile offspring

As they are **not** the same **species**, any offspring are infertile.

FOSSILS These are remains of plants and animals that have been preserved in rocks. A study of fossils tells us how life on earth has changed or evolved.

The fossil record is incomplete for many reasons:
• Soft tissue may not be fossilised.
• Fossils may not have been discovered.
• Fossilisation rarely occurs (animals and plants may be eaten or decomposed.

A narrowing indicates that many species died, reducing the total number. This suggests a mass **extinction**.

Here the **dinosaurs** become **extinct**. Evidence remains in the **fossil record**.

Few reptiles today.

The number of mammals is increasing.

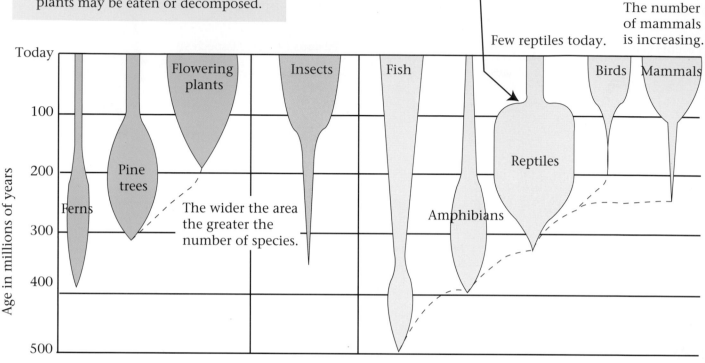

The wider the area the greater the number of species.

Algae and fungi present

Vertebrates evolved in the following order
Fish → Amphibians → Reptiles
↙ ↘
Mammals Birds
(evidence from the fossil record)

Plants and animals may die out or become **extinct** if conditions change, e.g. a change in climate, introduction of a predator, loss of habitat.
All of these might cause a species to become extinct.

For example, the **dodo** (extinct).
The dodo lived on the island of Mauritius in the Indian Ocean. It became extinct soon after European sailors arrived. The sailors introduced predators to the island such as pigs and dogs.
Between them, the dodo was hunted for food.
Being unable to fly and too large to hide there was no escape.

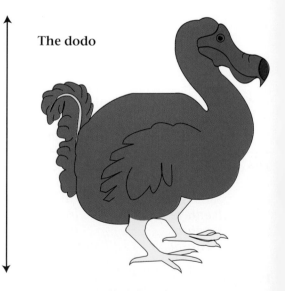

The dodo

1 metre

Questions:
1. What are fossils?
2. Where are fossils found?
3. Why is the fossil record incomplete?
4. Which groups were present around 400 million years ago?
5. What groups were dominant 200 million years ago, and how do we know?
6. Which two groups evolved from reptiles?
7. What does the term extinct mean?
8. Which bird became extinct in Mauritius, and why?

CLASSIFICATION OF LIVING THINGS Five kingdoms.

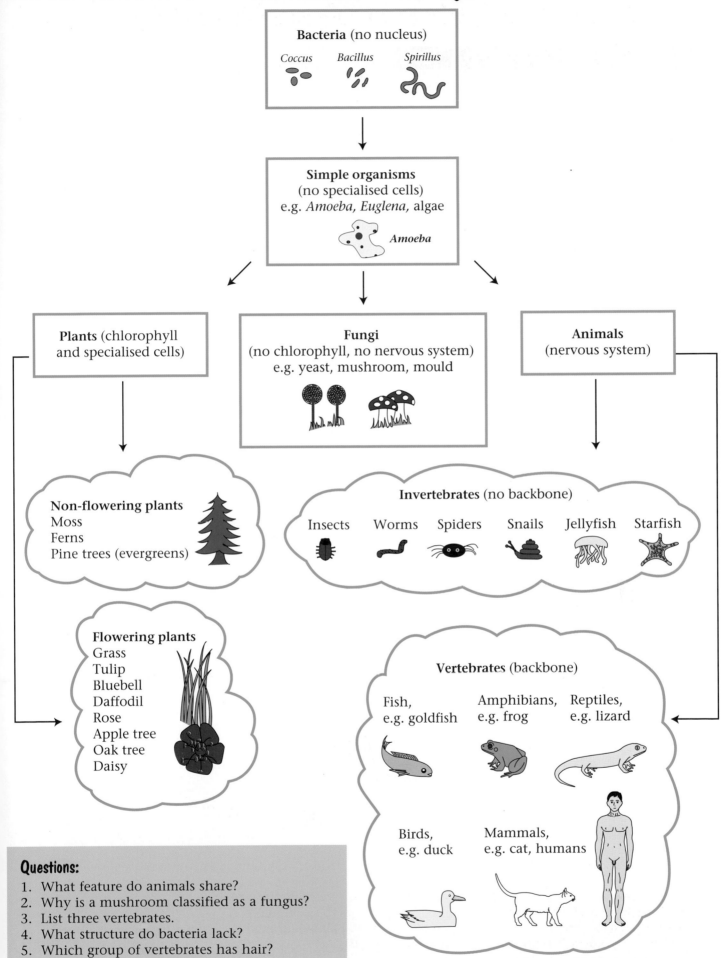

Bacteria (no nucleus)

Coccus   *Bacillus*   *Spirillus*

Simple organisms
(no specialised cells)
e.g. *Amoeba, Euglena*, algae

Amoeba

Plants (chlorophyll
and specialised cells)

Fungi
(no chlorophyll, no nervous system)
e.g. yeast, mushroom, mould

Animals
(nervous system)

Non-flowering plants
Moss
Ferns
Pine trees (evergreens)

Invertebrates (no backbone)

Insects   Worms   Spiders   Snails   Jellyfish   Starfish

Flowering plants
Grass
Tulip
Bluebell
Daffodil
Rose
Apple tree
Oak tree
Daisy

Vertebrates (backbone)

Fish,
e.g. goldfish

Amphibians,
e.g. frog

Reptiles,
e.g. lizard

Birds,
e.g. duck

Mammals,
e.g. cat, humans

Questions:
1. What feature do animals share?
2. Why is a mushroom classified as a fungus?
3. List three vertebrates.
4. What structure do bacteria lack?
5. Which group of vertebrates has hair?

ECOLOGY

FOOD CHAINS AND FOOD WEBS

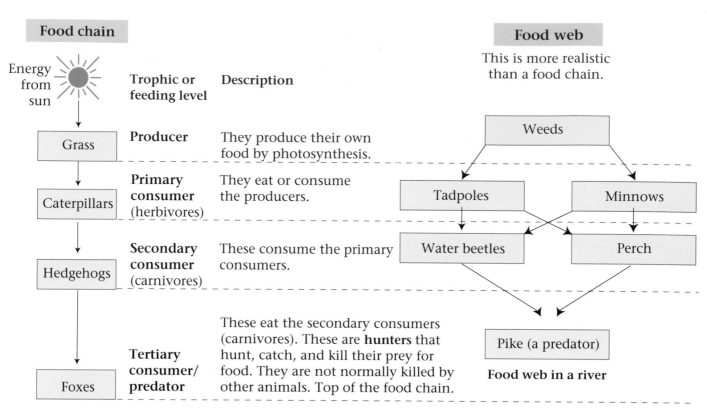

Food chain

Energy from sun

Trophic or feeding level	Description
Producer	They produce their own food by photosynthesis.
Primary consumer (herbivores)	They eat or consume the producers.
Secondary consumer (carnivores)	These consume the primary consumers.
Tertiary consumer/ predator	These eat the secondary consumers (carnivores). These are **hunters** that hunt, catch, and kill their prey for food. They are not normally killed by other animals. Top of the food chain.

Grass → Caterpillars → Hedgehogs → Foxes

A food chain is unrealistic as grass is eaten by many animals.
Foxes eat many animals not just hedgehogs.

Food web

This is more realistic than a food chain.

Weeds → Tadpoles, Minnows → Water beetles, Perch → Pike (a predator)

Food web in a river

Food web in a wood

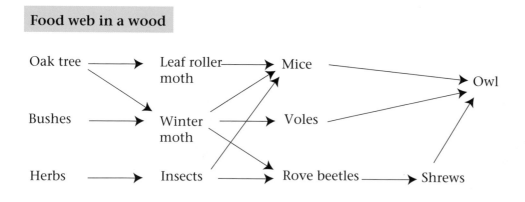

Oak tree → Leaf roller moth → Mice → Owl
Bushes → Winter moth → Voles
Herbs → Insects → Rove beetles → Shrews

Questions:

Use the food web above to answer the following questions:

1. Name a producer. Why is it called a producer?
2. What do mice eat?
3. What eats winter moths?
4. Which is a predator? How do you know?
5. Name two primary consumers.
6. Name two secondary consumers.
7. How many tertiary consumers are shown in this web?
8. Describe the possible effects on the food web if the voles died.
9. Which animal is found at two feeding levels?
10. Write out the longest food chain present in this web.

WOODLAND HABITAT An oak wood supports a varied community.

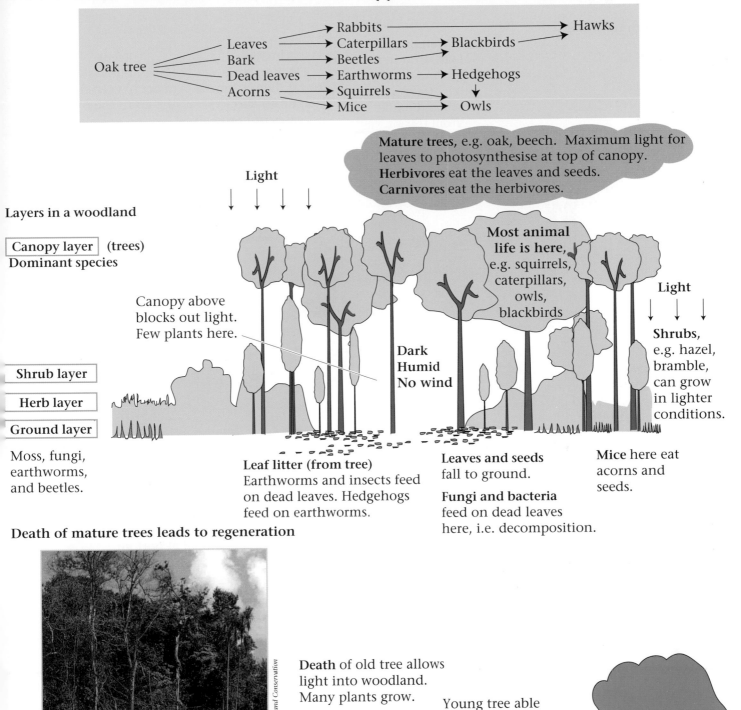

Oak tree
- Leaves → Rabbits → Hawks
- Leaves → Caterpillars → Blackbirds → Hawks
- Bark → Beetles → Blackbirds
- Dead leaves → Earthworms → Hedgehogs
- Acorns → Squirrels → Owls
- Acorns → Mice → Owls

Mature trees, e.g. oak, beech. Maximum light for leaves to photosynthesise at top of canopy. **Herbivores** eat the leaves and seeds. **Carnivores** eat the herbivores.

Light

Layers in a woodland

| Canopy layer | (trees)
Dominant species

Canopy above blocks out light. Few plants here.

Most animal life is here, e.g. squirrels, caterpillars, owls, blackbirds

Light

| Shrub layer |

| Herb layer |

| Ground layer |

Dark Humid No wind

Shrubs, e.g. hazel, bramble, can grow in lighter conditions.

Moss, fungi, earthworms, and beetles.

Leaf litter (from tree) Earthworms and insects feed on dead leaves. Hedgehogs feed on earthworms.

Leaves and seeds fall to ground.

Fungi and bacteria feed on dead leaves here, i.e. decomposition.

Mice here eat acorns and seeds.

Death of mature trees leads to regeneration

From Bowes: A Colour Atlas of Plant Propagation and Conservation

Fallen trees.

Death of old tree allows light into woodland. Many plants grow.

Young tree able to grow as **light** and space are present. As it gets bigger it will block out the light below.

Light

Questions:
Refer to the food web above:
1. What do squirrels eat?
2. What do blackbirds feed on?
3. Write out the longest food chain in the web.
4. If all the mice died how would it affect the food web?
5. Which organisms feed on dead leaves?

POND HABITAT How animals and plants are adapted to life in a pond.

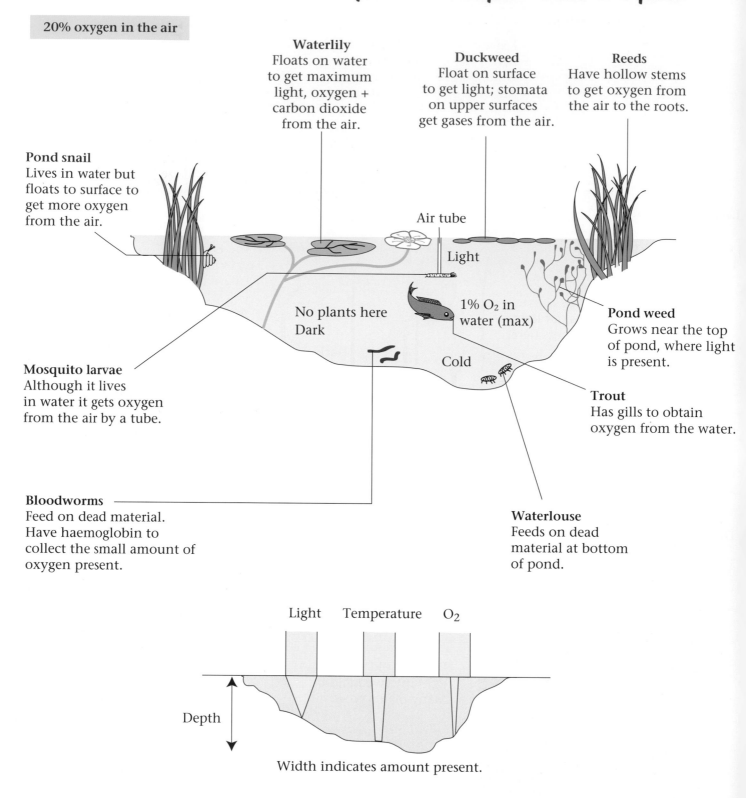

Waterlily
Floats on water to get maximum light, oxygen + carbon dioxide from the air.

Duckweed
Float on surface to get light; stomata on upper surfaces get gases from the air.

Reeds
Have hollow stems to get oxygen from the air to the roots.

Pond snail
Lives in water but floats to surface to get more oxygen from the air.

Air tube

Light

No plants here
Dark

1% O₂ in water (max)

Pond weed
Grows near the top of pond, where light is present.

Mosquito larvae
Although it lives in water it gets oxygen from the air by a tube.

Cold

Trout
Has gills to obtain oxygen from the water.

Bloodworms
Feed on dead material. Have haemoglobin to collect the small amount of oxygen present.

Waterlouse
Feeds on dead material at bottom of pond.

Light   Temperature   O₂

Depth

Width indicates amount present.

Questions:
1. How much oxygen is in water compared to the air?
2. Explain the distribution of plants.
3. What three factors decrease as you go deeper in the water?
4. Explain why mosquito larvae have an air tube.
5. Where are stomata usually found in aquatic plants? How does this compare with land plants?
6. How does haemoglobin help the bloodworm?
7. How do reeds get oxygen to their roots which are under water?

ANIMAL ADAPTATIONS

1. Animals in arctic conditions

e.g. **polar bear**,
reindeer, wolves,
Arctic fox, brown bears.

The **adaptations** of animals and plants determine their **distribution** and **abundance**.

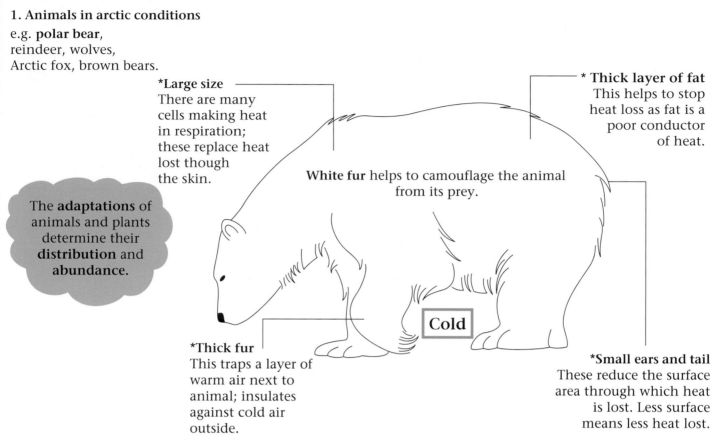

*Large size
There are many cells making heat in respiration; these replace heat lost though the skin.

* Thick layer of fat
This helps to stop heat loss as fat is a poor conductor of heat.

White fur helps to camouflage the animal from its prey.

Cold

*Thick fur
This traps a layer of warm air next to animal; insulates against cold air outside.

*Small ears and tail
These reduce the surface area through which heat is lost. Less surface means less heat lost.

* All these features are designed to **reduce heat loss** to prevent a fall in the animal's temperature in a cold climate.

2. Animals in deserts

e.g. **Fennec fox** (smallest fox), kangaroo rat, small desert fox, desert hedgehog.

*Light brown colour
Provides good camouflage against sand keeping it safe from predators.

*Small size
Few cells making heat prevents animal overheating. Also being small allows animals to burrow out of sun in hot day temperatures.

*Thin layer of fat
Heat is lost through this thin layer.

*Thin fur
Less warm air trapped. Allows heat to escape from the animal.

Hot

*Large ears and tail
This increases the surface area so more heat can be lost through the extra surface.

*All these features are designed to **increase heat loss** to prevent overheating in a hot climate.

Questions:
1. Animals found in arctic conditions are large. Can you suggest why?
2. How do large ears and tails help a mammal living in hot conditions?
3. Name 3 features found in mammals living in cold conditions. Explain why each is necessary.
4. How does thick fur help a mammal to keep warm in cold conditions?

ESTIMATING POPULATION SIZE

Animal populations

Most animals move making counting difficult.
The following equipment is used to collect and count
small animals.

Pooters
Pooters are used to suck small insects from
vegetation without harming them.

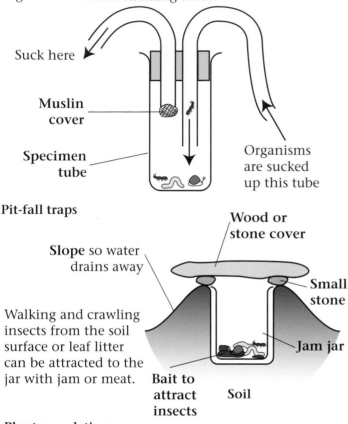

Suck here

Muslin cover

Specimen tube

Organisms
are sucked
up this tube

Pit-fall traps

Wood or
stone cover

Slope so water
drains away

Small stone

Walking and crawling
insects from the soil
surface or leaf litter
can be attracted to the
jar with jam or meat.

Jam jar

Bait to
attract
insects

Soil

Plant populations

Counting is simpler as plants do not move.
Hollow square frames called **quadrats** are used
to count plants.

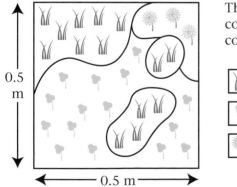

0.5 m

0.5 m

The plants can be
counted or the %
cover calculated

40% (grass)

50% (clover)

10% (daisies)

This is 0.25 m² quadrat frame commonly used in the
field.

Sampling
All these methods indicate what plants and animals
are present in a habitat. Normally only a few samples
are taken and the total numbers can then be estimated.
The more samples taken, the more reliable the estimate.

Nets e.g. sweep nets

A sweep net can be swept
through grass, bushes,
streams or ponds.
Small insects are
collected.

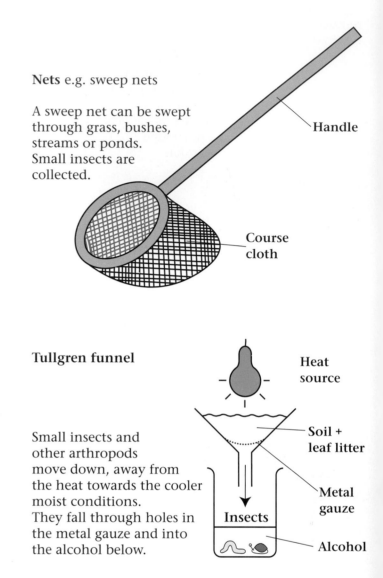

Handle

Course
cloth

Tullgren funnel

Heat
source

Soil +
leaf litter

Small insects and
other arthropods
move down, away from
the heat towards the cooler
moist conditions.
They fall through holes in
the metal gauze and into
the alcohol below.

Insects

Metal
gauze

Alcohol

Gridded quadrats can also be used.

There are 100 squares.

If a plant appears in
50 squares, this can be
called 50% occurrence.

Sampling must be repeated
many times at random to give
a reliable indication of the
plant population. The results
can then be averaged.
A random sampling method
avoids bias.

POPULATIONS

A population is a group of organisms of the same species in one area, e.g. human population in London.

Births and **immigration** *increase* the size of the population.

Immigration
↓
Births → Population → Deaths
↓
Emigration

Deaths and **emigration** *reduce* the size of the population.

Paramecium and yeast in a beaker
Paramecium eats yeast.

① *Paramecium* numbers rise as food is available.

④ Yeast numbers rise as fewer *Paramecium* are there to eat them.

② Yeast numbers fall as they are eaten.

③ *Paramecium* numbers fall as yeast is running out.

Population of cane toads in Queensland, Australia

Numbers rise when there is plenty of food, space and no predators, i.e. **no limiting factors.**

Rapid rise

Numbers increase slowly at start due to few toads present.

The number of cane toads is still rising.

Cane toads were introduced into Australia to feed on the caterpillars spoiling the sugar cane, an example of biological control. They were alien to the country and no initial tests were carried out. Unfortunately, the cane toads did not feed on the caterpillars and instead are feeding on other, often endangered, species. The number of cane toads is now rising out of control.

Predator–prey relationship:
Rabbits (prey) → Foxes (predators)

----Foxes
—Rabbits

One population controls the size of the other:

① Large numbers of rabbits provide food for foxes. So fox numbers increase, which causes rabbit numbers to decrease.
② Few rabbits mean little food for foxes, so fox numbers decrease.
③ Few foxes allows rabbit numbers to increase again.

Yeast population in a beaker

This is a **stable population,** staying steady.

Numbers steady as food and space is limited.

At start few yeast able to reproduce.

Rise in numbers as plenty of food, space and oxygen.

Questions:
1. What might cause the numbers in a population to increase?
2. What factors stop a population rising?
3. What is meant by a stable population?
4. How do rabbit numbers control fox numbers? What is this an example of?
5. What is a population?

53

PYRAMIDS OF NUMBERS Size of box depends on number (not mass).

The size of the box is determined by the **number** of organisms at each level.

One million e.g. grass $\longrightarrow$ 10 sheep $\longrightarrow$ 2 foxes

(Small number = small box)

(Secondary consumer)

(Primary consumer)

Fox | 2
Sheep | 10

Grass – one million   **(Producer)**

(Large number = large box)

Usually the size of the organism **increases** as you go **up**, but the **number decreases**.

Unusual pyramids of numbers

1. 1 oak tree $\longrightarrow$ 1,000 insects $\longrightarrow$ 20 sparrows $\longrightarrow$ 3 hawks

3 hawks

20 sparrows

1,000 insects

1 oak tree

The **size of a population** depends on how successfully it **competes** for factors such as: food, water, shelter, light, minerals.

Unusual pyramid of numbers

2. 500 potato plants $\longrightarrow$ 1 human $\longrightarrow$ 20 fleas $\longrightarrow$ 1,000 parasites

1,000 parasites on fleas

20 fleas

1 human

500 potato plants

Size increases

1 large organism eats

3 smaller ones which eat

15 smaller ones which eat many plants

Number decreases

Pyramids of number give no indication of the **mass** of each organism. One grass plant is given the same area as one oak tree in the pyramid. This is misleading.

Questions:
1. What determines the size of the box?
2. Why are there usually fewer secondary consumers than primary consumers?
3. How is a pyramid of numbers misleading?

PYRAMIDS OF BIOMASS Size of box depends on mass.

The size of the box is determined by the **mass** of organisms at each level.

Oak tree ⟶ Insects ⟶ Sparrows ⟶ Hawks

A pyramid of biomass

Mass of hawks

Mass of sparrows

Mass of insects

Mass of oak tree

The total weight of an organism in a particular area is called its biomass. Normally the mass decreases as you go up.

Pyramids of biomass are usually pyramid-shaped.

Mass of secondary consumer

Mass of primary consumer

Mass of producer

Dry mass is normally used as it is more accurate but it involves drying organisms, which kills them. Fresh mass is unreliable, especially in plants, as rain greatly increases mass for a while.

Problems
- Dry mass involves killing, which is undesirable.
- Mass varies at different times of year.
- Some masses are more productive, e.g. almost all of a grass plant produces sugar to feed herbivores; only a small percentage of an oak tree does this.

Questions:
1. What is meant by the term biomass?
2. What is dry mass and how is it measured?
3. What shape are pyramids of biomass?
4. Why does an oak tree produce less sugar per gram than grass?

PYRAMIDS OF ENERGY

The size of each box in the food chain below is determined by the **energy** flowing through each level. (Energy is measured in **kilojoules – kJ**.)

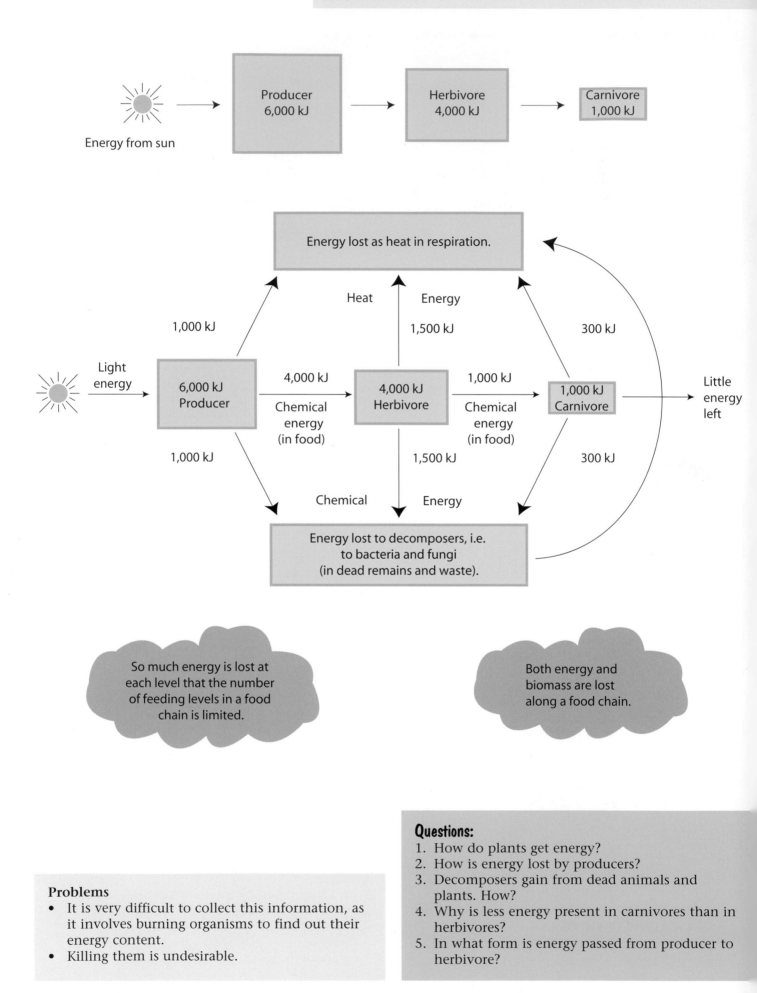

So much energy is lost at each level that the number of feeding levels in a food chain is limited.

Both energy and biomass are lost along a food chain.

Problems
- It is very difficult to collect this information, as it involves burning organisms to find out their energy content.
- Killing them is undesirable.

Questions:
1. How do plants get energy?
2. How is energy lost by producers?
3. Decomposers gain from dead animals and plants. How?
4. Why is less energy present in carnivores than in herbivores?
5. In what form is energy passed from producer to herbivore?

ENERGY LOSSES AND FOOD PRODUCTION

Energy enters the food chain from the sun. When a cow eats grass, the energy passes to the cow. A great deal of energy is lost at each feeding level.

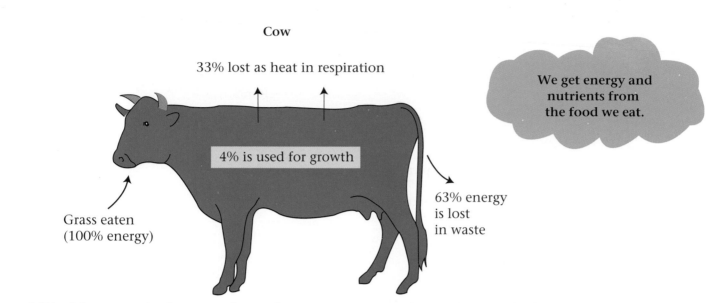

Cow

33% lost as heat in respiration

4% is used for growth

63% energy is lost in waste

Grass eaten (100% energy)

We get energy and nutrients from the food we eat.

96% of the energy in the grass taken in by a cow is **not** passed on to the next feeding level.

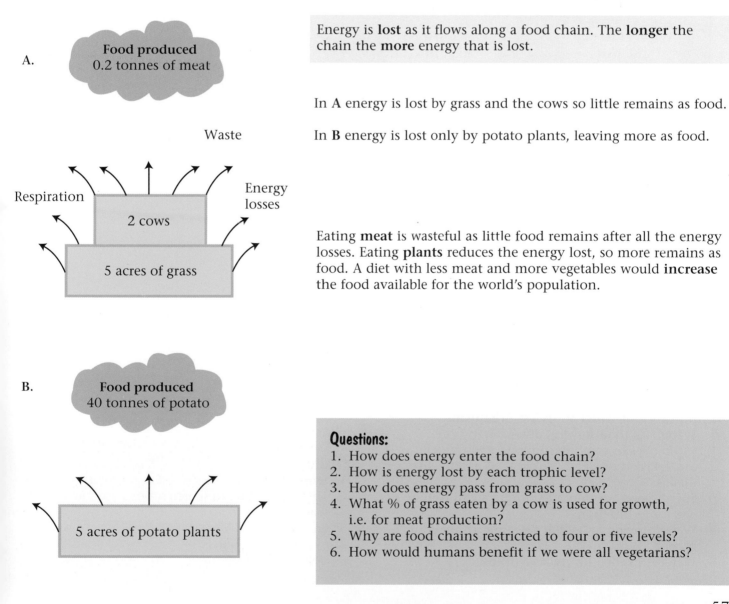

A.

Food produced 0.2 tonnes of meat

Energy is **lost** as it flows along a food chain. The **longer** the chain the **more** energy that is lost.

In **A** energy is lost by grass and the cows so little remains as food.

In **B** energy is lost only by potato plants, leaving more as food.

Waste

Respiration

Energy losses

2 cows

5 acres of grass

Eating **meat** is wasteful as little food remains after all the energy losses. Eating **plants** reduces the energy lost, so more remains as food. A diet with less meat and more vegetables would **increase** the food available for the world's population.

B.

Food produced 40 tonnes of potato

5 acres of potato plants

Questions:
1. How does energy enter the food chain?
2. How is energy lost by each trophic level?
3. How does energy pass from grass to cow?
4. What % of grass eaten by a cow is used for growth, i.e. for meat production?
5. Why are food chains restricted to four or five levels?
6. How would humans benefit if we were all vegetarians?

WATER CYCLE

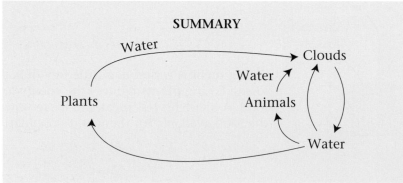

Importance of water to living organisms
1. Water is a good solvent, able to transport substances in solution in blood and phloem.
2. Evaporation of water (in sweat) is an effective means of cooling.
3. Sperm need a watery medium to swim to the egg for fertilisation.
4. Water provides a habitat for some animals and plants.
5. Water is needed for photosynthesis.

Questions:
1. By what process does water enter plant roots?
2. Name two ways in which animals produce water.
3. How do plants lose water?
4. What forms clouds?
5. What is precipitation?

CARBON CYCLE Carbon is present in protein, carbohydrate and lipid.

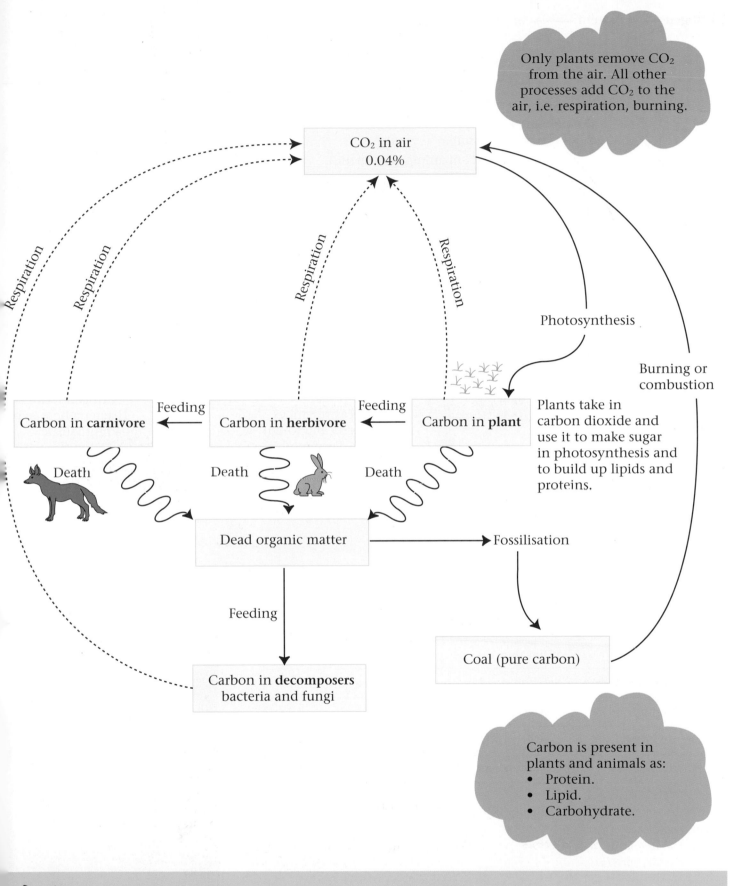

Only plants remove CO_2 from the air. All other processes add CO_2 to the air, i.e. respiration, burning.

CO_2 in air 0.04%

Respiration

Photosynthesis

Burning or combustion

Feeding

Carbon in **carnivore**

Carbon in **herbivore**

Carbon in **plant**

Plants take in carbon dioxide and use it to make sugar in photosynthesis and to build up lipids and proteins.

Death

Dead organic matter

Fossilisation

Feeding

Coal (pure carbon)

Carbon in **decomposers** bacteria and fungi

Carbon is present in plants and animals as:
• Protein.
• Lipid.
• Carbohydrate.

Questions:
1. For what process do plants take in CO_2?
2. What do plants use the CO_2 to make?
3. What process adds CO_2 to the air by all living organisms?
4. How does carbon pass from plants to animals?
5. What feeds on dead animals and plants?
6. How do the decomposers return carbon to the air?
7. What happens to plants when they are fossilised? How does this cause CO_2 to be added to the air?

NITROGEN CYCLE

Most plants cannot use the **nitrogen** in the air as it is **insoluble**. **Nitrates** are and enter roots dissolved in water.

Nitrogen ⟶ Protein ⟶ Growth

All living things need **nitrogen** to make **protein** for **growth**.

N₂ air 79%

Nitrogen-fixing bacteria in root nodules, e.g. of **mutualism**.

Nitrates

Pea plant

N₂ fixing bacteria

Sugar

Denitrifying bacteria (remove nitrates from soil).

(N in animal as protein.)

Feeding

Animal protein

Feeding

Lightning puts nitrates into the soil

N is used to make plant protein.

Death

Faeces and urine (contains N₂)

Death

Death

Soluble nitrates taken up by plant roots.

Dead organic matter (contains N₂)

Death

Provides plant with nitrates

Nitrates

Feeding

Root nodules
Nitrogen-fixing bacteria here have a mutualistic relationship with the plant as they both benefit from the relationship.

Decay microbes (bacteria and fungi) are **decomposers**

Waste product

Pea family (leguminous plant).

Nitrifying bacteria (change ammonia to nitrates).

Ammonia (NH₃)

Bacteria in nitrogen cycle	
Type	**What they do**
Saprobiotic (decay)	Feed on dead material and release ammonia
Nitrifying	Convert ammonia to nitrates
Denitrifying	Convert nitrates to nitrogen gas
Nitrogen fixing	Convert nitrogen gas to nitrates

From Bowes: *A Colour Atlas of Plant Structure*

Root system of Soybean showing root nodules. The nitrogen-fixing bacteria that live here provide the plant with nitrates.

Questions:
1. Why do plants and animals need nitrogen?
2. How do plants get their nitrogen? How does this nitrogen enter plants?
3. How much nitrogen is in the air?
4. Why can't most plants make use of the nitrogen in the air?
5. What process passes nitrogen into animals from plants?
6. What are decomposers?
7. What substance is produced by the decomposers as a waste product?
8. Which bacteria produce nitrates in the soil?

HUMAN EFFECTS ON THE ENVIRONMENT
IMPORTANCE OF TROPICAL RAIN FORESTS

These are found in hot and wet areas, e.g. South America, Western Africa, Indonesia, Australia, and South-East Asia. Many trees are being cut down to provide land for **farming** and **housing**. The large scale, permanent removal of forests is called **deforestation**.

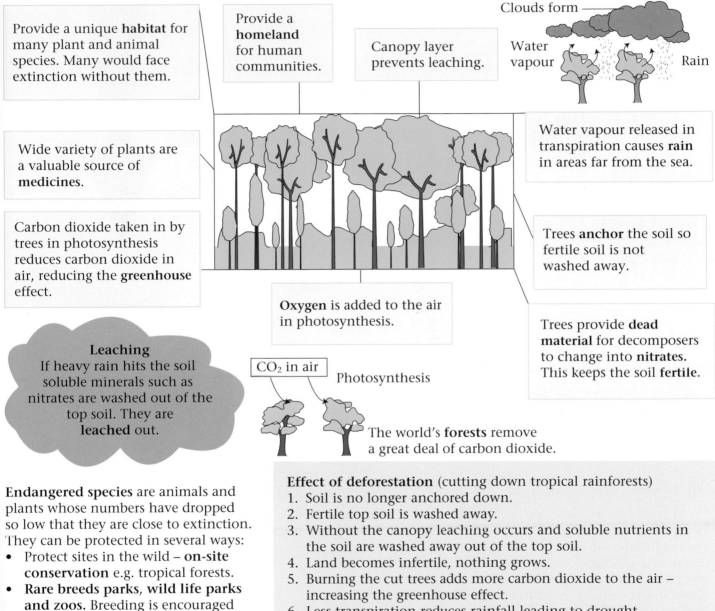

Provide a unique **habitat** for many plant and animal species. Many would face extinction without them.

Provide a **homeland** for human communities.

Canopy layer prevents leaching.

Clouds form

Water vapour — Rain

Wide variety of plants are a valuable source of **medicines**.

Water vapour released in transpiration causes **rain** in areas far from the sea.

Carbon dioxide taken in by trees in photosynthesis reduces carbon dioxide in air, reducing the **greenhouse** effect.

Trees **anchor** the soil so fertile soil is not washed away.

Oxygen is added to the air in photosynthesis.

Trees provide **dead material** for decomposers to change into **nitrates**. This keeps the soil **fertile**.

Leaching
If heavy rain hits the soil soluble minerals such as nitrates are washed out of the top soil. They are **leached** out.

CO_2 in air — Photosynthesis

The world's **forests** remove a great deal of carbon dioxide.

Endangered species are animals and plants whose numbers have dropped so low that they are close to extinction. They can be protected in several ways:
- Protect sites in the wild – **on-site conservation** e.g. tropical forests.
- **Rare breeds parks, wild life parks and zoos.** Breeding is encouraged and predators removed.
- **Seed banks** – prevent the loss of genes in plants.
- **Legal protection** of certain species.
- **Educate** the public.

Effect of deforestation (cutting down tropical rainforests)
1. Soil is no longer anchored down.
2. Fertile top soil is washed away.
3. Without the canopy leaching occurs and soluble nutrients in the soil are washed away out of the top soil.
4. Land becomes infertile, nothing grows.
5. Burning the cut trees adds more carbon dioxide to the air – increasing the greenhouse effect.
6. Less transpiration reduces rainfall leading to drought.
7. Less trees means less photosynthesis and more carbon $dioxide_2$ in the air, adding to the greenhouse effect.
8. Loss of habitats has caused some organisms extinction, others are endangered.
9. In the short term the increase in dead material results in more decomposers respiring, so releasing more carbon dioxide into the air.

Questions:
1. Why does cutting down the trees cause the extinction of some animals?
2. How do trees help to reduce the greenhouse effect?
3. Where are tropical rain forests found in the world? Name two places.
4. How is rain produced by trees?
5. Without trees, the land may become infertile. How?

THE GREENHOUSE EFFECT This increases global warming.

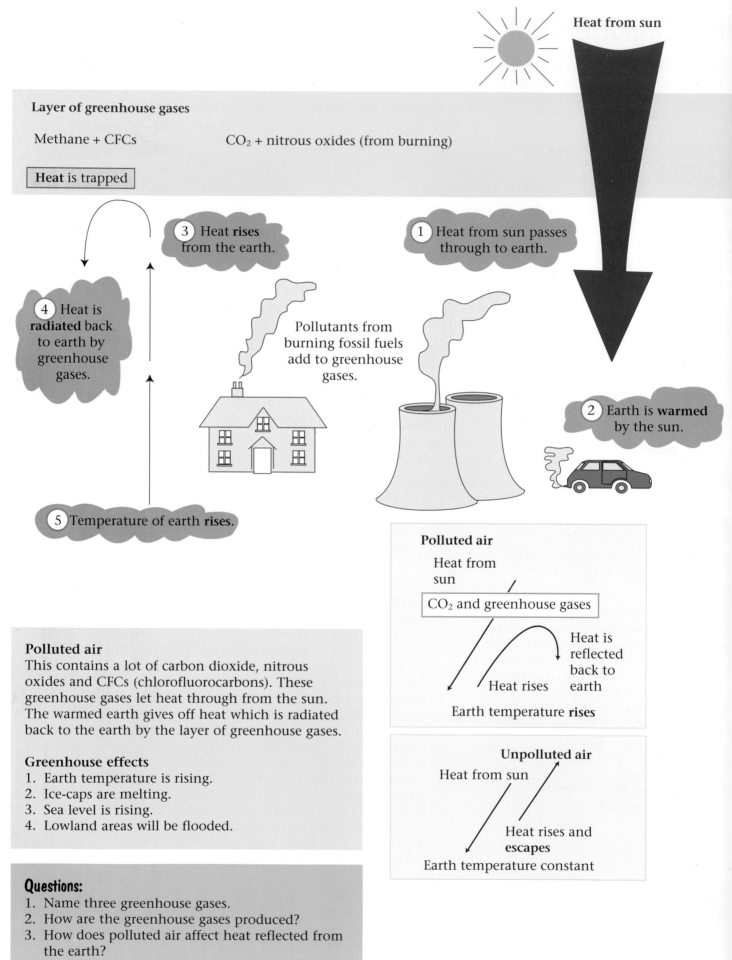

Heat from sun

Layer of greenhouse gases

Methane + CFCs CO_2 + nitrous oxides (from burning)

Heat is trapped

3 Heat **rises** from the earth.

1 Heat from sun passes through to earth.

4 Heat is **radiated** back to earth by greenhouse gases.

Pollutants from burning fossil fuels add to greenhouse gases.

2 Earth is **warmed** by the sun.

5 Temperature of earth **rises**.

Polluted air

Heat from sun

CO_2 and greenhouse gases

Heat is reflected back to earth

Heat rises

Earth temperature **rises**

Unpolluted air

Heat from sun

Heat rises and **escapes**

Earth temperature constant

Polluted air
This contains a lot of carbon dioxide, nitrous oxides and CFCs (chlorofluorocarbons). These greenhouse gases let heat through from the sun. The warmed earth gives off heat which is radiated back to the earth by the layer of greenhouse gases.

Greenhouse effects
1. Earth temperature is rising.
2. Ice-caps are melting.
3. Sea level is rising.
4. Lowland areas will be flooded.

Questions:
1. Name three greenhouse gases.
2. How are the greenhouse gases produced?
3. How does polluted air affect heat reflected from the earth?
4. Name two effects of global warming.

AIR POLLUTION

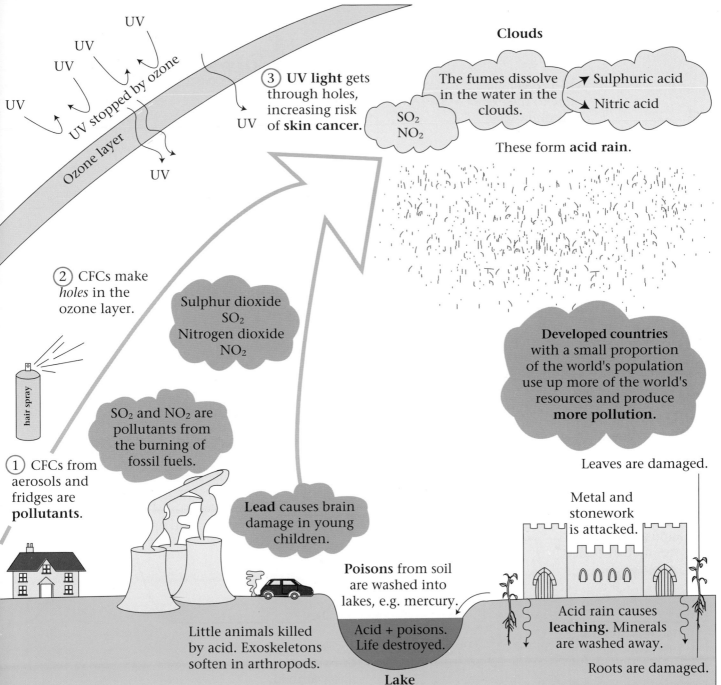

UV
UV
UV
UV
UV

UV stopped by ozone

Ozone layer

UV

UV

③ **UV light** gets through holes, increasing risk of **skin cancer.**

Clouds

The fumes dissolve in the water in the clouds.

➤ Sulphuric acid
▲ Nitric acid

SO_2
NO_2

These form **acid rain.**

② CFCs make *holes* in the ozone layer.

Sulphur dioxide
SO_2
Nitrogen dioxide
NO_2

hair spray

SO_2 and NO_2 are pollutants from the burning of fossil fuels.

① CFCs from aerosols and fridges are **pollutants.**

Developed countries with a small proportion of the world's population use up more of the world's resources and produce **more pollution.**

Lead causes brain damage in young children.

Leaves are damaged.

Metal and stonework is attacked.

Poisons from soil are washed into lakes, e.g. mercury.

Little animals killed by acid. Exoskeletons soften in arthropods.

Acid + poisons. Life destroyed.

Acid rain causes **leaching.** Minerals are washed away.

Roots are damaged.

Lake

Prevention of acid rain
The amount of sulphur dioxide and nitrogen oxides produced in burning can be reduced if we:
1. Use a catalytic converter in cars to reduce the pollutants released.
2. Design furnaces so that some pollutants are trapped, e.g. a wet scrubber removes sulphur dioxide.
3. Clean coal and oil before burning so that less sulphur is present.

CFC = chlorofluorocarbons

Lichens as indicator species
Lichens are unusual organisms, as they are a mixture of two different species, algae and fungi. They form the crusty pale surface on trees in rural areas. Also the yellow powder and black 'paint' looking crusts on rocky shores. Lichens are indicators of air quality. If SO_2 levels are low, lichens are abundant. With rising SO_2 levels, lichens are scarce.

Questions:
1. List three effects of acid rain.
2. How does the ozone layer protect humans?
3. How do chlorofluorocarbons affect the ozone layer?
4. Why is soil often infertile in acid rain regions?
5. How does lead from burning fuels affect human health?

POLLUTION IN A RIVER

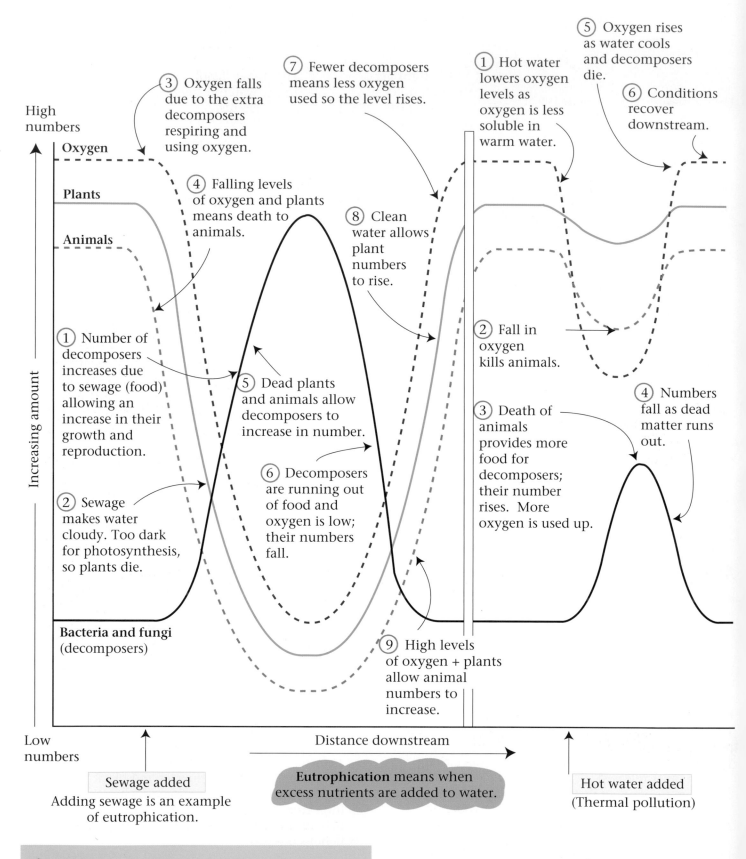

③ Oxygen falls due to the extra decomposers respiring and using oxygen.

⑦ Fewer decomposers means less oxygen used so the level rises.

① Hot water lowers oxygen levels as oxygen is less soluble in warm water.

⑤ Oxygen rises as water cools and decomposers die.

⑥ Conditions recover downstream.

High numbers

Oxygen

Plants

Animals

④ Falling levels of oxygen and plants means death to animals.

⑧ Clean water allows plant numbers to rise.

Increasing amount

① Number of decomposers increases due to sewage (food) allowing an increase in their growth and reproduction.

⑤ Dead plants and animals allow decomposers to increase in number.

② Fall in oxygen kills animals.

④ Numbers fall as dead matter runs out.

② Sewage makes water cloudy. Too dark for photosynthesis, so plants die.

⑥ Decomposers are running out of food and oxygen is low; their numbers fall.

③ Death of animals provides more food for decomposers; their number rises. More oxygen is used up.

Bacteria and fungi (decomposers)

⑨ High levels of oxygen + plants allow animal numbers to increase.

Low numbers

Distance downstream

Sewage added

Adding sewage is an example of eutrophication.

Eutrophication means when excess nutrients are added to water.

Hot water added (Thermal pollution)

Questions:
1. Why do the numbers of bacteria and fungi rise when sewage is added?
2. What gas are these decomposers taking in for respiration?
6. What is eutrophication?
7. Explain why oxygen levels fall when hot water is added to a river?

Indicator species are organisms whose presence indicates particular levels of oxygen or pH in the water. For example, the presence of trout indicates high levels of oxygen.

EUTROPHICATION The sudden increase in the nutrient content of a lake or river

e.g. when excess nitrates are washed out of the soil and into a lake (leaching).

Farmer adds nitrate fertiliser to farmland. Nitrates increase plant growth and profit.

Rain pours on the soil and soluble nitrates are washed into a lake.

Nitrates in the lake cause the rapid growth of algae on the lake surface.

A thick 'blanket' of algae covers the lake surface and blocks light to the plants below. This is an **algal bloom**.

The blanket of algae die due to their short life cycle.

There is an increase in the dead material in the lake due to the dead surface algae and the plants below.

The **decomposers**, bacteria and fungi, feed on the dead material and their number increases.

Decomposers are **aerobic**, using up oxygen in respiration. Large numbers of decomposers respiring, reduce the oxygen level of the lake.

Oxygen level in the lake falls.

Aerobic organisms (those requiring oxygen) die.

Dead lake

Courtesy of Paul Glendell

Effects of eutrophication. Dead fish caused by release of sewage into a river. This resulted in increased decomposers and a fall in oxygen levels.

Questions:
1. Why do farmers add nitrates to the soil?
2. What causes the nitrate to pass into rivers and lakes?
3. How does nitrate cause an algal bloom and what is it?
4. Algae only live for a short time. What feeds on the dead algae?
5. What happens to the oxygen level in the lake and why?
6. Which organisms are affected by the change in oxygen level.
7. What does the word eutrophication mean?
8. Can eutrophication be prevented?

FISH FARMING A managed ecosystem (aquaculture).

The North Sea was a rich fishing area with plenty of cod, haddock, herring, plaice, and mackerel. Overfishing has now reduced the supplies and fishing restrictions have been brought in to protect fish numbers in two main ways:

- Restricting the numbers of fish caught; a quota system.
- Only catching larger fish. Increasing the mesh size allows smaller fish to escape and live long enough to breed.

Fishermen furious at planned restrictions
A 40,000 square mile area of the North Sea, almost $1/5$ of its entire area may be off limits to cod, haddock and whiting fishermen as part of a desperate attempt to ensure survival of the cod stock.

SALMON fish farming developed during the 1970s to provide large quantities of salmon, without depleting the wild salmon numbers. By 1980, 800 tonnes of salmon was produced on the west coast of Scotland, providing a rich source of protein. The salmon are farmed in cages in sea lochs and in sheltered inlets of the sea, where conditions can be controlled. Although huge numbers are farmed, pollution problems are causing serious concerns.

A wild salmon leaping up a river to reach the pool where it hatched out. Here it will mate and produce more salmon.

EUTROPHICATION
Faeces from the fish and uneaten food, fall to the bottom of the loch, causing an increase in the number of decomposers that feed on it. The decomposers use up oxygen so oxygen levels fall on the sea bottom, causing death to organisms here.

CHEMICAL POLLUTION
A large number of salmon in a confined space are more likely to have and to spread disease. The fish louse, which feeds on the fish, is widespread and is treated with the chemical **dichlorvos**. This chemical affects other organisms in the food chain killing lobsters, prawns, and other fish.

The pollution and parasites from the salmon farms have resulted in a dramatic fall in the number of wild salmon. Wild salmon have been found with up to 500 lice feeding on them.

Questions:
1. Why does fishing need to be regulated?
2. In what two ways are fish numbers being protected?
3. When and where did salmon fish farming develop?
4. What problems are caused by the vast numbers of salmon kept in cages?
5. What effect have farmed salmon had on wild salmon?

FARMED SALMON – Stages in production.

Newly hatched
salmon in fresh water
hatchery

Controlled food
provided

Controlled light, with many longer periods of
light mimicking summers, speed up growth of fish.

70 m diameter
(250,000 fish in here)

Cage for
salmon

Fish moved to **cages**
in sea loch

Fish packed tight, up to 250,000 in each cage.
Disease and parasites a problem, e.g. sea lice.

Pesticides
used to kill lice

Chemical seeps
out to rest of sea

People expect salmon
to be pink. In the wild
this is due to salmon
eating pink crustaceans,
such as prawn and shrimp.
Farmed salmon is given
pink pigment in the diet,
adding about 15% to
the cost of food.

Salmon **grow** quickly

Salmon fed pellets of concentrated small fish,
vaccines, growth hormones, and pink colouring.

By 2000 the West Coast
of Scotland produced 127,000
tonnes of salmon.

Salmon starved for
10 days

This empties the gut so
increasing the storage life of
the fish.

Pollution from the fish
farms here is comparable
to the sewage from
9.4 million people.

Salmon reach
the shops

If salmon farms were
situated in areas with
good tidal flow, most
of these problems would
have been prevented.

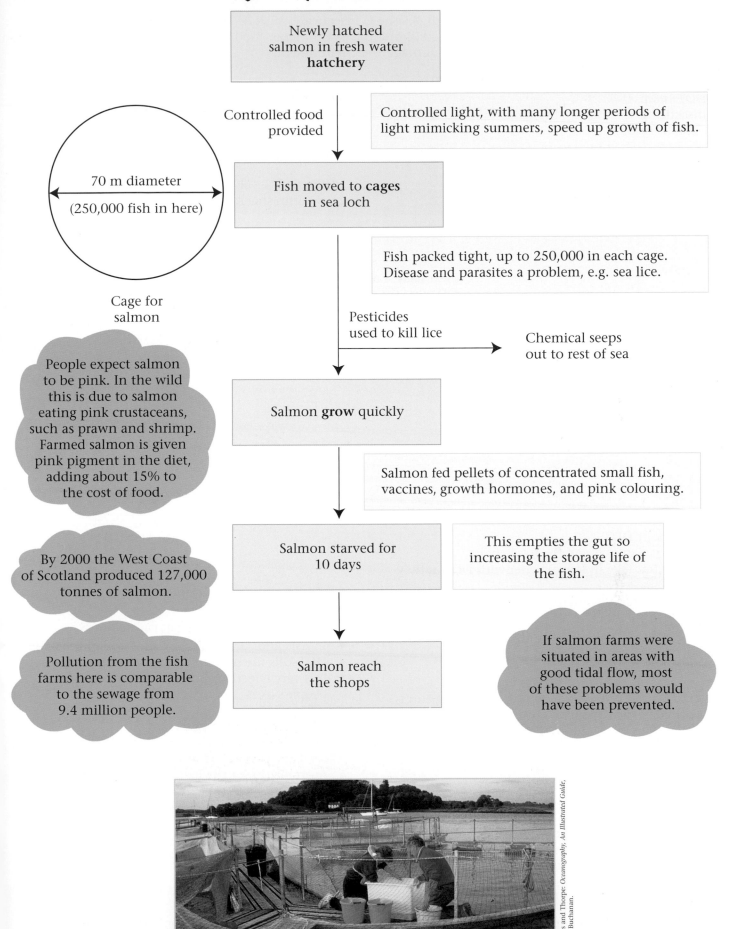

From Summerhayes and Thorpe: *Oceanography: An Illustrated Guide*,
courtesy of Dr Jim Buchanan.

Salmon farming.

PESTICIDES Poisonous chemicals which kill pests.

Pesticides are poisonous chemicals which kill pests, e.g. insecticides such as DDT kill insects. Fungicides kill fungi. Pesticides are used to kill weeds and to kill caterpillars eating our food crops. These can **improve** our crop production but unfortunately they can also **damage** the environment.

DDT is an insecticide – it kills insects

DDT in ppm

Plankton ⟶ Fish ⟶ Grebes
(0.04)   (2)   (25)

ppm = parts per million

Pesticides like DDT are harmful to wildlife as they do not break down. Unchanged they pass up the food chain causing death.

High DDT levels cause bird eggs to have thin shells and break easily.

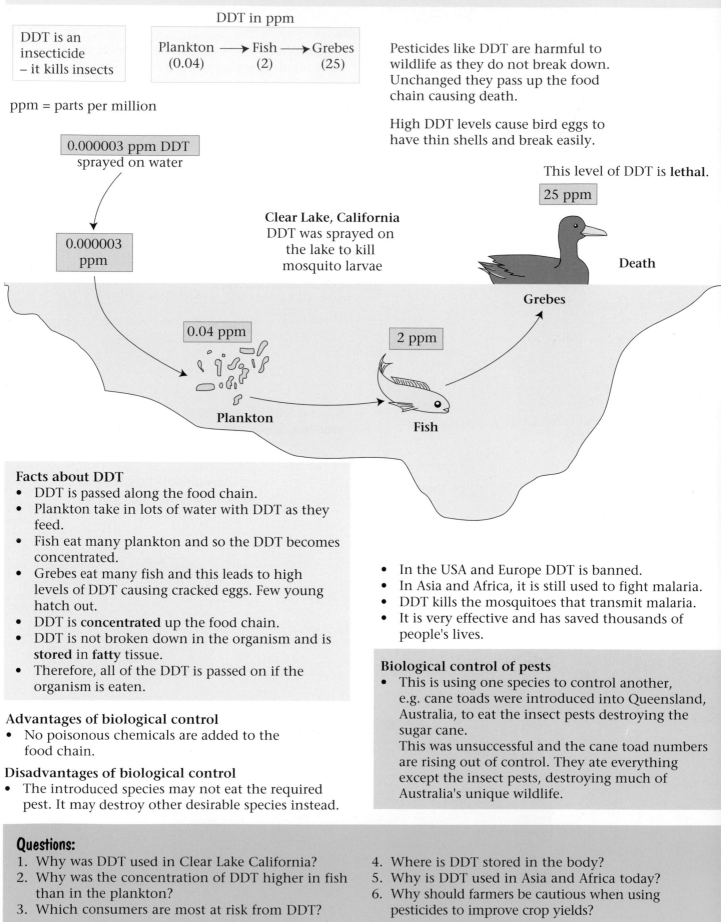

0.000003 ppm DDT
sprayed on water

0.000003 ppm

Clear Lake, California
DDT was sprayed on the lake to kill mosquito larvae

This level of DDT is **lethal**.

25 ppm

Death

Grebes

0.04 ppm

2 ppm

Plankton

Fish

Facts about DDT
- DDT is passed along the food chain.
- Plankton take in lots of water with DDT as they feed.
- Fish eat many plankton and so the DDT becomes concentrated.
- Grebes eat many fish and this leads to high levels of DDT causing cracked eggs. Few young hatch out.
- DDT is **concentrated** up the food chain.
- DDT is not broken down in the organism and is **stored** in **fatty** tissue.
- Therefore, all of the DDT is passed on if the organism is eaten.

Advantages of biological control
- No poisonous chemicals are added to the food chain.

Disadvantages of biological control
- The introduced species may not eat the required pest. It may destroy other desirable species instead.

- In the USA and Europe DDT is banned.
- In Asia and Africa, it is still used to fight malaria.
- DDT kills the mosquitoes that transmit malaria.
- It is very effective and has saved thousands of people's lives.

Biological control of pests
- This is using one species to control another, e.g. cane toads were introduced into Queensland, Australia, to eat the insect pests destroying the sugar cane.
  This was unsuccessful and the cane toad numbers are rising out of control. They ate everything except the insect pests, destroying much of Australia's unique wildlife.

Questions:
1. Why was DDT used in Clear Lake California?
2. Why was the concentration of DDT higher in fish than in the plankton?
3. Which consumers are most at risk from DDT?
4. Where is DDT stored in the body?
5. Why is DDT used in Asia and Africa today?
6. Why should farmers be cautious when using pesticides to improve crop yields?

MICROBES

USEFUL AND HARMFUL MICROBES

	Bacteria	Fungi	Viruses
Structure	*Coccus Bacillus Spirillus*	Spore case — Hyphae threads	Protective coat — Genetic material (DNA)
Useful	• Yoghurt making • Breaking down dead material to release useful substances, e.g. nitrates • Making insulin in genetic engineering • Breaking down sewage • Making vinegar	• Producing penicillin (an antibiotic) • Cheese making • Yeast is used in baking and wine making • Mushrooms provide food • Fungi are used to make mycoprotein – a food source	• Used in genetic engineering to make useful products
Harmful	**Cause diseases** • Tuberculosis • Food poisoning *(Salmonella)* • Tonsillitis • Whooping cough • Tetanus Bacteria feed on our food and cause **decay**	**Cause diseases** • Athlete's foot • Ringworm Fungi feed on our food and cause **decay**	**Cause diseases** • Influenza • Chicken pox • Small pox • Mumps • Rabies • Polio • AIDS
Examples	• *Bacillus tuberculosi* • *Salmonella*	• Penicillin • Bread mould • Yeast • Mushroom	• HIV(human immunodeficiency virus) The AIDS virus attacks the white blood cells which help our bodies fight disease.

Antibiotics

Antibiotics are chemicals used to kill bacteria. Penicillin was the first antibiotic discovered, in London, by Sir Alexander Fleming in 1928. It was a huge breakthrough in the fight against bacterial diseases.

> *Find out how Fleming's research led to this important discovery.*

Over use of antibiotics can lead to the evolution of resistant bacteria, such as MRSA, which has caused the death of many hospital patients.

Questions:
1. How are fungi useful to us? Name two ways.
2. Name three diseases caused by viruses.
3. Give three examples of fungi.
4. Which microbe is made of hyphae threads?

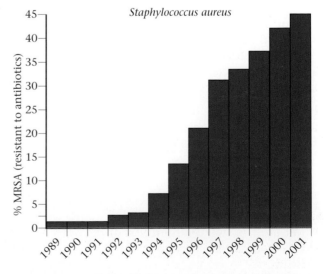

The percentage of *Staphylococcus aureus* resistant to methicillin (MRSA), in England and Wales 1989–2001.

DECOMPOSERS The decay organisms.

There are two main groups, **bacteria** and **fungi**. They feed on dead material and return useful nutrients back into the soil to be used again, e.g. nitrates.

Bacteria and fungi are called **Saprobiotic** as they feed on **dead**, organic matter, such as dead animals and plants. They also may cause decay in our food.

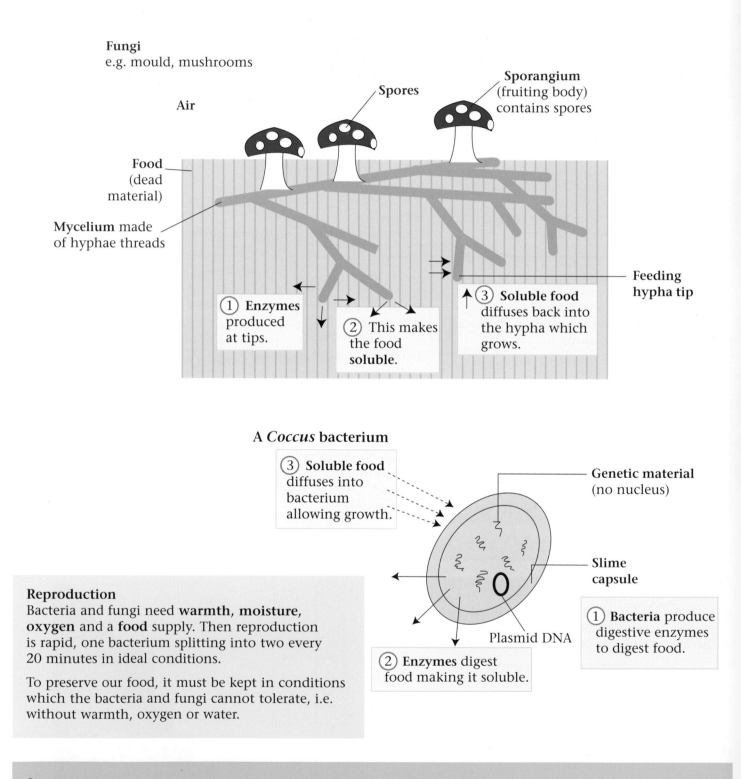

Fungi
e.g. mould, mushrooms

Air

Spores

Sporangium
(fruiting body)
contains spores

Food
(dead
material)

Mycelium made
of hyphae threads

Feeding
hypha tip

① **Enzymes**
produced
at tips.

② This makes
the food
soluble.

③ **Soluble food**
diffuses back into
the hypha which
grows.

A *Coccus* bacterium

③ **Soluble food**
diffuses into
bacterium
allowing growth.

Genetic material
(no nucleus)

Slime
capsule

Plasmid DNA

② **Enzymes** digest
food making it soluble.

① **Bacteria** produce
digestive enzymes
to digest food.

Reproduction
Bacteria and fungi need **warmth, moisture, oxygen** and a **food** supply. Then reproduction is rapid, one bacterium splitting into two every 20 minutes in ideal conditions.

To preserve our food, it must be kept in conditions which the bacteria and fungi cannot tolerate, i.e. without warmth, oxygen or water.

Questions:
1. What are the two main groups of decomposers?
2. How often can bacteria reproduce in ideal conditions?
3. What causes our food to decay?
4. What are the three types of bacteria?
5. What would happen if there were no decomposers?

TREATMENT OF SEWAGE

Grid
Breaks up sewage and stops entry of rubbish.

Sewer

Raw sewage
Faeces, urine, waste water, rain water.

Sewage → Methane gas
Sewage → Fertiliser
Sewage → Clean water

Sludge tank

Liquid continues

~Sludge~

Organic solids settle as sludge which is removed.

Sludge out

Settlement tank
Grit and heavy solids settle here.

Here sludge is broken down by **anaerobic bacteria**. Digested sludge is dried to produce **fertiliser. Methane gas** is also produced.

Liquid sprinkled on stones. Rotation adds oxygen for **aerobic bacteria.**

Dried

Sludge digested by **anaerobic bacteria.**

O_2 O_2

Fertiliser

Methane gas (a fuel)

Digester

Stones
covered in **aerobic bacteria** and **protozoa** that feed on organic material. Bacteria destroy remaining organic material making it safe.

River

Cleaned water added to river.

Two types of bacteria help to break down sewage:
a) **Aerobic** bacteria (need oxygen).
b) **Anaerobic** bacteria (do not need oxygen).

Bacteria are used to break down sewage to clean water, methane gas and fertiliser. The **aerobic** bacteria feed on the organic material in the liquid sewage, resulting in clean water. The solid organic material, sludge, is digested by the **anaerobic** bacteria, producing **fertiliser** and **methane gas**, both useful products.

Problems caused by an increasing human population

- increased consumption of limited resources, e.g. coal, oil, wood
- build-up of greenhouse gases
- sewage disposal
- air and water pollution
- loss of habitats leading to loss of species, i.e. reduced biodiversity
- waste disposal

Recycling of glass, tins and paper is now routine in most areas of the UK. The use of biodegradable materials is also on the rise.

Questions:
1. Which two types of bacteria are needed to break down sewage?
2. Why is sewage treatment necessary?
3. What useful products are gained from sewage treatment?

FOOD PRESERVATION

Food goes **bad** because **bacteria** and **fungi** feed on it causing **decay**. Bacteria and fungi need **warmth**, **oxygen**, and **water**. If any of these are removed the food will stay fresh as microbes cannot survive without these conditions.

Example	Type	What is done	How it works	How long it stays fresh
Baked beans	Tinned foods	Sealed in airtight container. Heated to a high temperature (no oxygen).	No oxygen present. High temperature destroys microbes.	For years, unless the tins are damaged or punctured.
Milk Margarine	Fridge	Kept at 2°C (low temperature).	Cool temperature slows down growth and reproduction of microbes. Slows rate of decay.	For days. When the fridge is opened, warm air enters, allowing reproduction of microbes.
Ice-cream Peas	Frozen foods	Kept at -18°C (no warmth).	Very low temperature stops growth and reproduction but it does not kill microbes. No decay.	For years. Food will come out with the same quantity of microbes as went into the freezer.
Dried potato Dried milk	Dried foods	All the water is removed (no water).	No water means no decay.	For years, unless packet is opened. Damp air will then allow entry and growth of microbes.
Jam Pickled onions Bacon	Chemicals	Placed in chemicals. **Salt:** bacon, ham (no water).	Salt **dries** the food, so no water, no decay.	Weeks.
		Sugar: marmalade, jam (removes water).	Sugar **dries** microbes.	Months, if unopened.
		Vinegar: pickles, chutney (acid conditions).	Acid **destroys** bacteria and fungi.	Months.

Food labelling

Processed food often contains a high proportion of fat and salt.

Too much salt in the diet can lead to increased blood pressure for about 30% of the population.

> *Choose 3 highly processed foods, e.g. baked beans, crisps and pot noodles. Write down the energy content of each and list all the additives present.*

> *Use the internet to find out why these additives are added.*

Questions:
1. What conditions do bacteria and fungi need?
2. How does canning foods stop decay?
3. What effect does freezing have on the microbes in food?
4. Why does bacon last for several weeks without going bad?
5. Pickles last for months. How does their treatment preserve them?

HOW DISEASES SPREAD

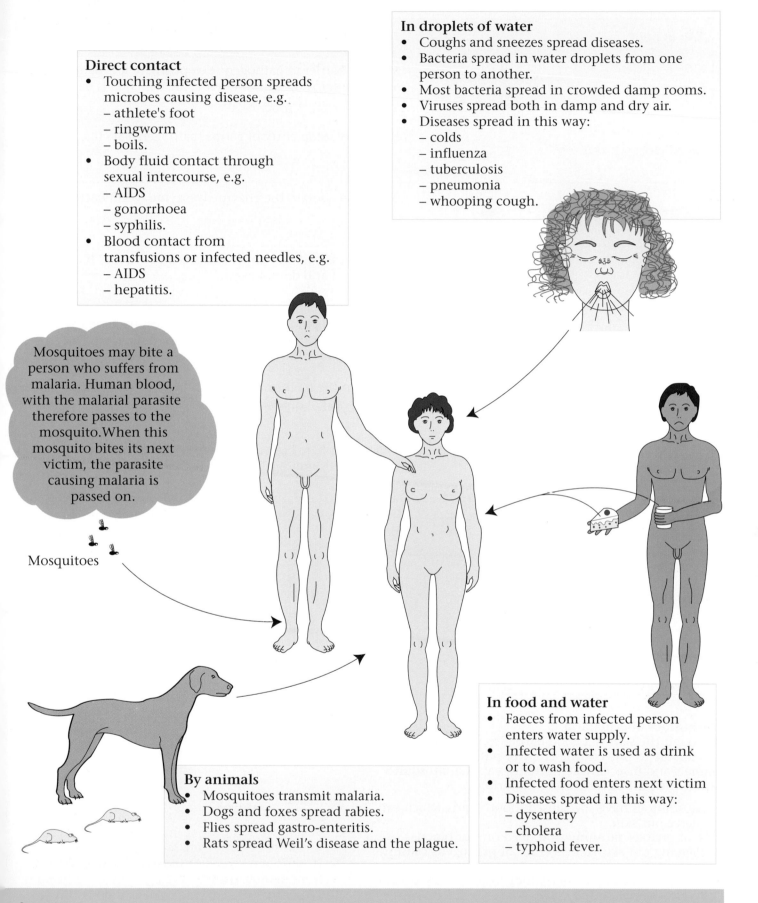

Direct contact
- Touching infected person spreads microbes causing disease, e.g.
  – athlete's foot
  – ringworm
  – boils.
- Body fluid contact through sexual intercourse, e.g.
  – AIDS
  – gonorrhoea
  – syphilis.
- Blood contact from transfusions or infected needles, e.g.
  – AIDS
  – hepatitis.

In droplets of water
- Coughs and sneezes spread diseases.
- Bacteria spread in water droplets from one person to another.
- Most bacteria spread in crowded damp rooms.
- Viruses spread both in damp and dry air.
- Diseases spread in this way:
  – colds
  – influenza
  – tuberculosis
  – pneumonia
  – whooping cough.

Mosquitoes may bite a person who suffers from malaria. Human blood, with the malarial parasite therefore passes to the mosquito. When this mosquito bites its next victim, the parasite causing malaria is passed on.

Mosquitoes

By animals
- Mosquitoes transmit malaria.
- Dogs and foxes spread rabies.
- Flies spread gastro-enteritis.
- Rats spread Weil's disease and the plague.

In food and water
- Faeces from infected person enters water supply.
- Infected water is used as drink or to wash food.
- Infected food enters next victim
- Diseases spread in this way:
  – dysentery
  – cholera
  – typhoid fever.

Questions:
1. How is malaria spread?
2. What diseases are spread if sewage is not properly treated?
3. Why should someone with whooping cough keep away from healthy people?
4. In what conditions do bacterial infections spread most rapidly? How can this be prevented?

DEFENCES OF THE BODY TO PATHOGENS

A **pathogen** is an organism that causes disease.

Part of body		How it works
Skin		Skin is a barrier to pathogens except where it is damaged. Sweat contains the enzyme lysozyme which kills bacteria by breaking their cell wall open.
Blood clot	Platelet	Blood clots stop entry of germs (pathogens) at cuts.
Eyes (tears)		Tears also contain the enzyme lysozyme to kill bacteria.
Senses	Nose (smell) Tongue (taste)	Food that is bad may smell or taste 'off'. We find it unpleasant and do not eat it.
Stomach	Acid	Strong hydrochloric acid (pH2) kills most bacteria that are present in our food. Contaminated food may make us vomit, which will remove the harmful bacteria, and food.
Respiratory passages		Mucus in the trachea and bronchi traps bacteria, and cilia hairs sweep them away from the lungs.
Lungs (alveoli)		Phagocytes here keep the surface of the alveoli clean by surrounding and digesting any bacteria present.
Phagocytes		These white cells in the blood surround and digest invading bacteria.
Lymphocytes		Lymphocytes in the blood produce antibodies to kill the germs and antitoxins to destroy their poisons.
Reproductive system		Little defence here. Vagina and urethra (in penis) have acid conditions which destroys some bacteria.

Mosquitoes and malaria

Although the skin is a good barrier to most pathogens, it is unable to prevent the entry of **pathogens** from animals and insects which **bite** us, e.g. the **mosquito** carries the **pathogen** causing **malaria**.

- Mosquitoes have specialised mouth parts which can pierce our skin.
- This enables mosquitoes to feed on our blood.
- Piercing our skin allows entry of pathogens causing malaria.
- Mosquitoes can pass on pathogens from one person to another when they bite.
- Malaria can be spread quickly and easily by mosquitoes.

If travelling to malarial regions of the world, it is essential to take anti-malaria tablets and use mosquito nets and creams to prevent their bites.

Questions:

1. How do blood clots help to stop disease?
2. What feature of the stomach helps to prevent the entry of pathogens?
3. How do our senses of smell and taste protect us from pathogens?
4. What does the enzyme lysozyme do and where is it produced?
5. Why do mosquitoes bite us?
6. What disease can mosquitoes spread?
7. How does the mucus in the breathing tubes help to protect us from pathogens?

ANTIBODIES AND IMMUNITY

Disease is caused by invading bacteria and viruses. These are **pathogens** (disease-causing organisms). White blood cells called **lymphocytes** produce antibodies which are specific to the pathogen.

Antibodies attack both the organism and the poisons (toxins) they produce. Antibodies that attack toxins are called antitoxins

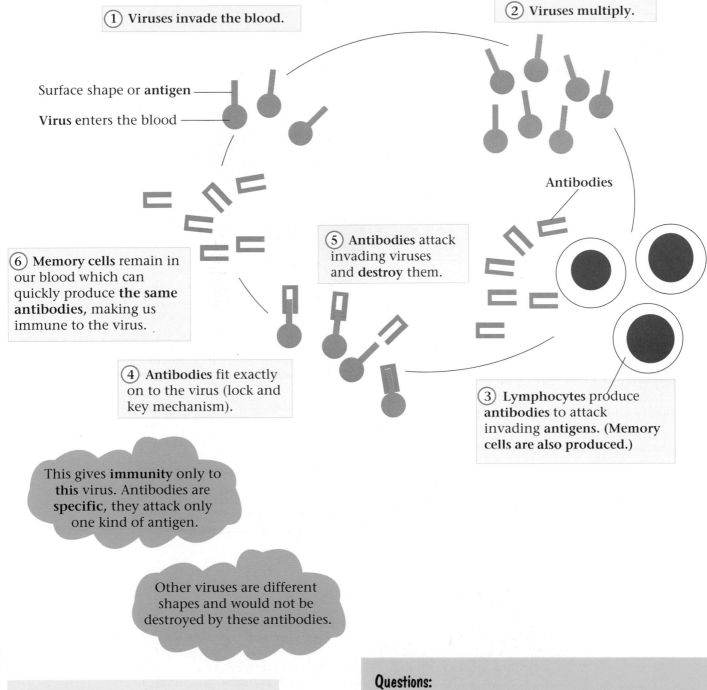

(1) Viruses invade the blood.

(2) Viruses multiply.

Surface shape or **antigen**

Virus enters the blood

Antibodies

(5) Antibodies attack invading viruses and **destroy** them.

(6) Memory cells remain in our blood which can quickly produce **the same antibodies**, making us immune to the virus.

(4) Antibodies fit exactly on to the virus (lock and key mechanism).

(3) Lymphocytes produce **antibodies** to attack invading **antigens. (Memory cells are also produced.)**

This gives **immunity** only to **this** virus. Antibodies are **specific**, they attack only one kind of antigen.

Other viruses are different shapes and would not be destroyed by these antibodies.

Vaccination
Weakened or dead microorganisms – the **vaccine** may be injected into a patient. The vaccine stimulates the patient's white blood cells to secrete antibodies specific to the vaccine, i.e. of matching shape. The patient becomes resistant or immune to that microorganism without being ill.

Questions:
1. Which white blood cells produce antibodies?
2. How can viruses enter the body?
3. What happens to viruses once inside the body?
4. What do we mean by an antigen?
5. Why are antibodies a particular shape? What does this achieve?
6. How do antibodies make us immune to a particular virus?
7. Why don't antibodies give us immunity to all invading viruses?
8. What is meant by a pathogen?

THE HUMAN BODY

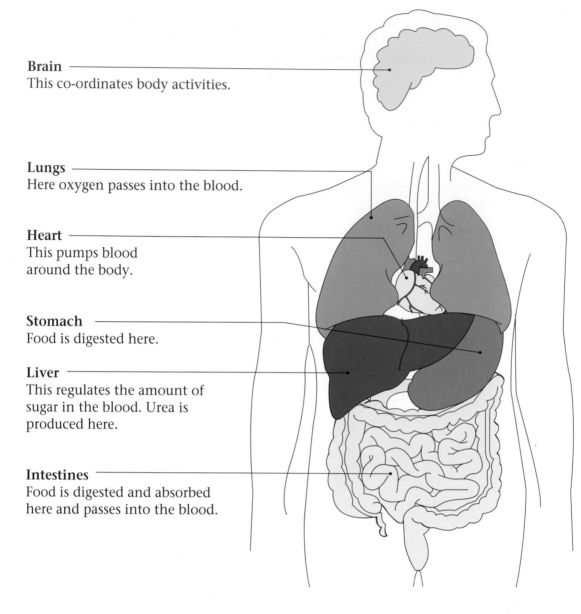

Brain
This co-ordinates body activities.

Lungs
Here oxygen passes into the blood.

Heart
This pumps blood
around the body.

Stomach
Food is digested here.

Liver
This regulates the amount of
sugar in the blood. Urea is
produced here.

Intestines
Food is digested and absorbed
here and passes into the blood.

The human body is divided into 3 parts:

Part	Organs present
Head	Brain, eyes, ears
Thorax	Lungs, heart
Abdomen	Liver, stomach, intestines, kidneys (only visible when intestines removed), reproductive organs (not shown)

A sheet of muscle called the diaphragm separates the thorax from the abdomen (see pages 79 and 90).

Questions:
1. In which organs is food digested?
2. Where does oxygen pass into the blood?
3. Which two major organs are found in the thorax?
4. Moving from the head to the abdomen, list the organs that are visible in the diagram.

NUTRITION AND CIRCULATION
HUMAN TEETH

Humans have two sets of teeth. The first set are called **milk teeth** and these are replaced by **permanent teeth** from the age of five.

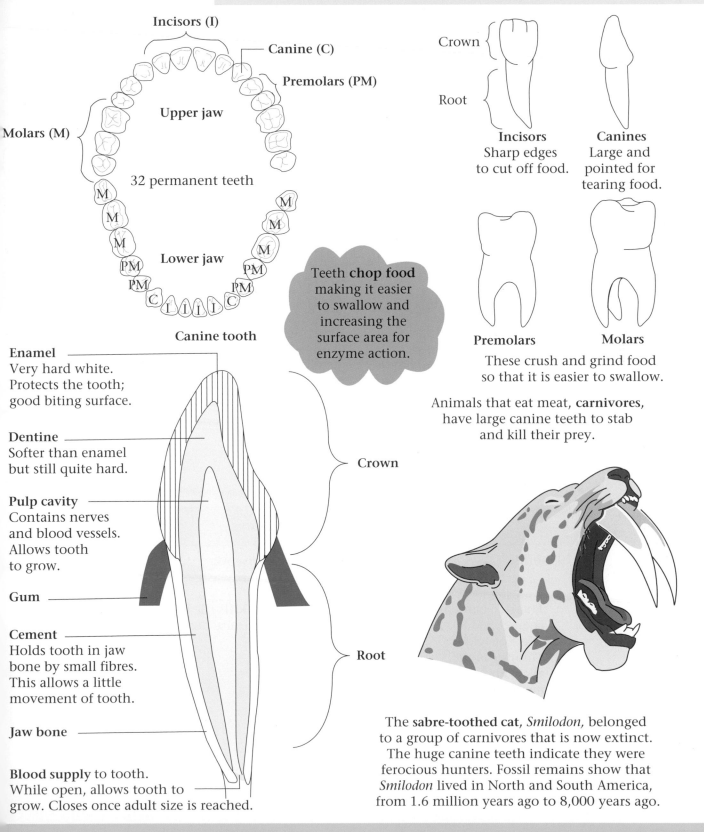

Incisors (I)

Canine (C)

Premolars (PM)

Upper jaw

Molars (M)

32 permanent teeth

Lower jaw

M M M M PM PM PM PM C C I I I I

Canine tooth

Enamel
Very hard white.
Protects the tooth;
good biting surface.

Dentine
Softer than enamel
but still quite hard.

Pulp cavity
Contains nerves
and blood vessels.
Allows tooth
to grow.

Gum

Cement
Holds tooth in jaw
bone by small fibres.
This allows a little
movement of tooth.

Jaw bone

Blood supply to tooth.
While open, allows tooth to
grow. Closes once adult size is reached.

Crown

Root

Crown

Root

Incisors
Sharp edges
to cut off food.

Canines
Large and
pointed for
tearing food.

Teeth **chop food** making it easier to swallow and increasing the surface area for enzyme action.

Premolars   **Molars**
These crush and grind food
so that it is easier to swallow.

Animals that eat meat, **carnivores**,
have large canine teeth to stab
and kill their prey.

The **sabre-toothed cat**, *Smilodon*, belonged
to a group of carnivores that is now extinct.
The huge canine teeth indicate they were
ferocious hunters. Fossil remains show that
Smilodon lived in North and South America,
from 1.6 million years ago to 8,000 years ago.

Questions:
1. What are the four types of teeth?
2. What protects our teeth from damage?
3. Why do molars and premolars have rough edges?
4. Which teeth are pointed?
5. Why do sheep not have canine teeth?
6. Herbivores like sheep spend a lot of time chewing. Why do their teeth continue to grow all their life?
7. Which part of the tooth contains nerves and blood?
8. What do incisors do?

TOOTH DECAY

After eating, sugar is left in the mouth. Bacteria feed on the sugar and produce an **acid** that causes decay. The white, sticky mixture of food and bacteria is called **plaque**. This should be removed regularly to avoid tooth decay and damage to the gums.

Stages in tooth decay

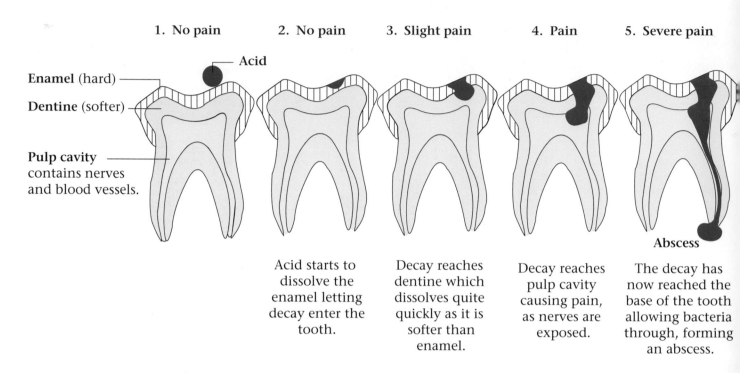

1. No pain	2. No pain	3. Slight pain	4. Pain	5. Severe pain

Enamel (hard)

Dentine (softer)

Pulp cavity contains nerves and blood vessels.

Acid

Abscess

	Acid starts to dissolve the enamel letting decay enter the tooth.	Decay reaches dentine which dissolves quite quickly as it is softer than enamel.	Decay reaches pulp cavity causing pain, as nerves are exposed.	The decay has now reached the base of the tooth allowing bacteria through, forming an abscess.

Prevention of tooth decay – five ways				
Diet	Fluoride	Brushing teeth	Antiseptic mouthwash	Dentist
Calcium and **Vitamin D** harden teeth, giving greater resistance to decay.	1. **Fluoride** hardens enamel, reducing effect of acid. 2. **Fluoride** is an **alkali** and neutralises the acid causing decay.	This **removes** trapped food which is a source of sugar for bacteria.	This **kills** the bacteria that cause decay.	**Dentists** can identify holes caused by decay and fill them, thus preventing further entry of bacteria.

Questions:
1. What causes tooth decay?
2. Which part of the tooth is affected by acid first?
3. When will pain first appear and why?
4. Why should the dentist be visited before pain starts?
5. How can tooth decay be prevented?

THE HUMAN DIGESTIVE SYSTEM (I)

This is where the digestion and absorption of food takes place.

Human digestive system

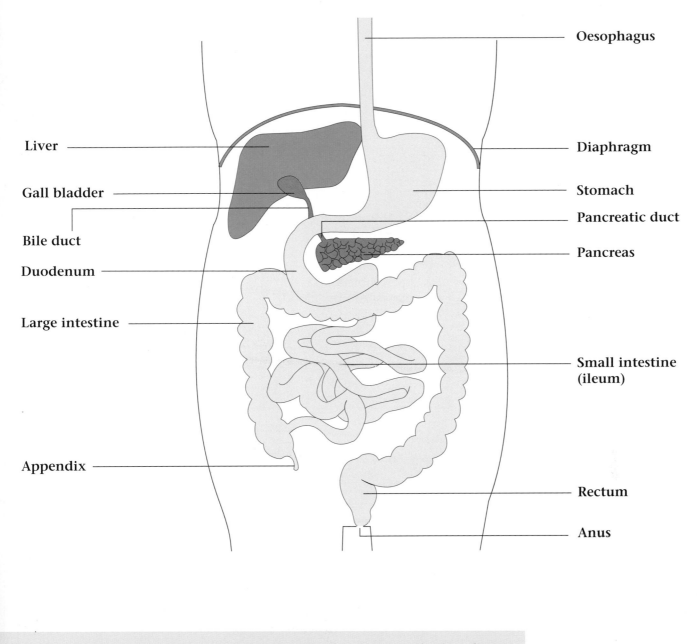

- Oesophagus
- Liver
- Gall bladder
- Bile duct
- Duodenum
- Large intestine
- Appendix
- Diaphragm
- Stomach
- Pancreatic duct
- Pancreas
- Small intestine (ileum)
- Rectum
- Anus

Digestion is the process in which **large, insoluble** substances are broken down to **small soluble** substances, which can be **absorbed** (pass through to blood).

Large, insoluble food $\xrightarrow{\text{Digestion}}$ Small, soluble food

Questions:
1. What are the tubes that food passes through from the mouth to anus? List them in the correct order.
2. Where is the gall bladder found?
3. What separates the thorax from the abdomen?
4. Why does food need to be digested?

THE HUMAN DIGESTIVE SYSTEM (II)

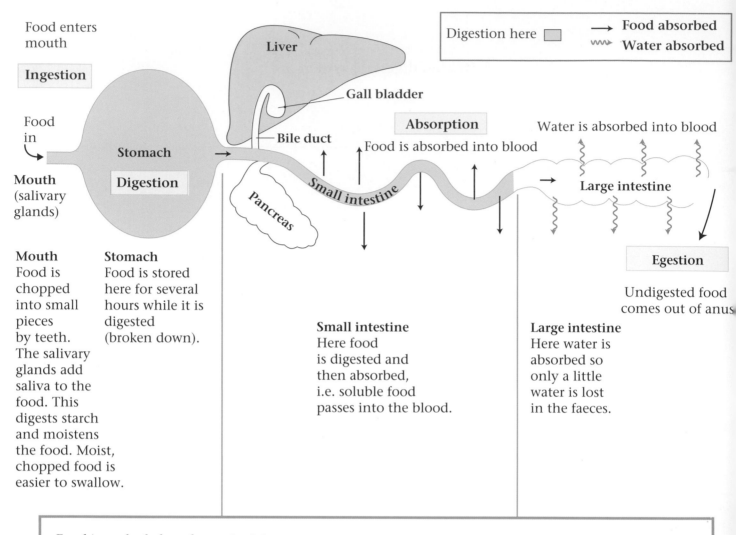

Food enters mouth

Ingestion

Food in

Mouth (salivary glands)

Stomach

Digestion

Liver

Gall bladder

Bile duct

Pancreas

Small intestine

Absorption

Food is absorbed into blood

Water is absorbed into blood

Large intestine

Egestion

Undigested food comes out of anus

Digestion here [ ] → Food absorbed ⟿ Water absorbed

Mouth
Food is chopped into small pieces by teeth. The salivary glands add saliva to the food. This digests starch and moistens the food. Moist, chopped food is easier to swallow.

Stomach
Food is stored here for several hours while it is digested (broken down).

Small intestine
Here food is digested and then absorbed, i.e. soluble food passes into the blood.

Large intestine
Here water is absorbed so only a little water is lost in the faeces.

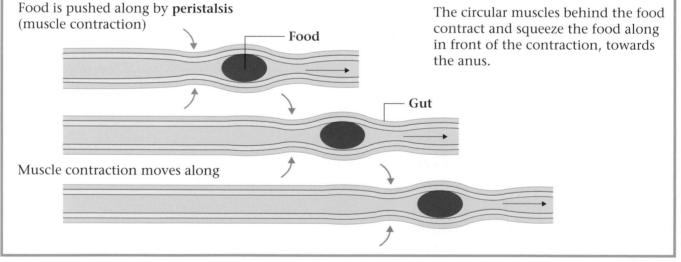

Food is pushed along by **peristalsis** (muscle contraction)

Food

Gut

Muscle contraction moves along

The circular muscles behind the food contract and squeeze the food along in front of the contraction, towards the anus.

DUODENUM

Bile • Emulsifies fats.
• Neutralises acid.
Pancreas - secretes enzymes:
• protease
• amylase
• lipase.

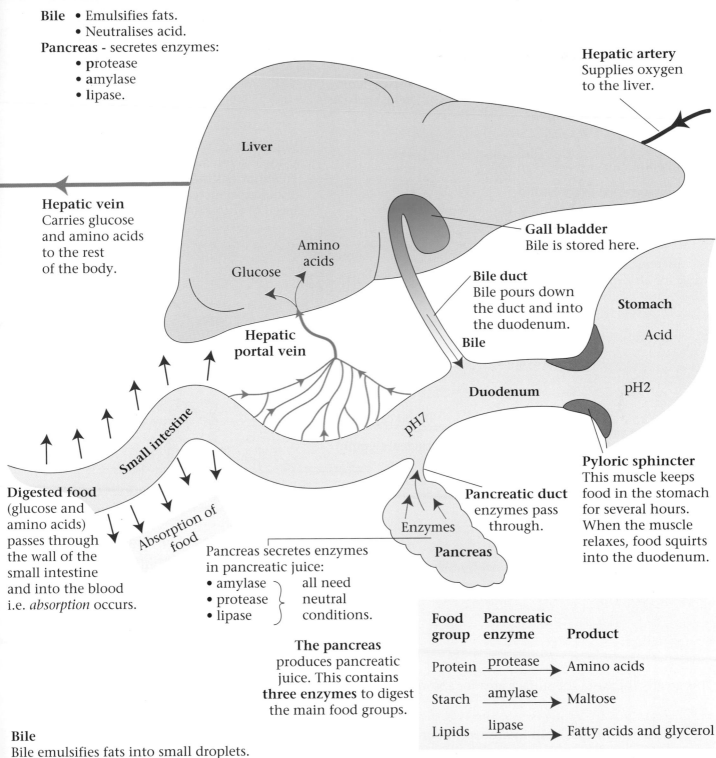

Hepatic artery
Supplies oxygen to the liver.

Liver

Hepatic vein
Carries glucose and amino acids to the rest of the body.

Amino acids

Glucose

Hepatic portal vein

Gall bladder
Bile is stored here.

Bile duct
Bile pours down the duct and into the duodenum.
Bile

Stomach
Acid

pH2

Duodenum

pH7

Small intestine

Digested food
(glucose and amino acids) passes through the wall of the small intestine and into the blood i.e. *absorption* occurs.

Absorption of food

Pyloric sphincter
This muscle keeps food in the stomach for several hours. When the muscle relaxes, food squirts into the duodenum.

Pancreatic duct
enzymes pass through.

Enzymes

Pancreas

Pancreas secretes enzymes in pancreatic juice:
• amylase ⎫
• protease ⎬ all need neutral conditions.
• lipase ⎭

The pancreas produces pancreatic juice. This contains **three enzymes** to digest the main food groups.

Food group	Pancreatic enzyme	Product
Protein	protease →	Amino acids
Starch	amylase →	Maltose
Lipids	lipase →	Fatty acids and glycerol

Bile
Bile emulsifies fats into small droplets.

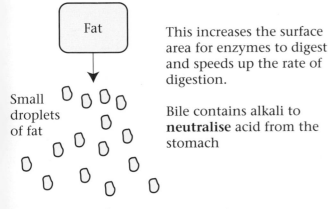

Fat

Small droplets of fat

This increases the surface area for enzymes to digest and speeds up the rate of digestion.

Bile contains alkali to **neutralise** acid from the stomach

Questions:
1. What are the three main food groups?
2. What enzymes are released by the pancreas?
3. What pH do the pancreatic enzymes need?
4. What are the two main functions of bile?
5. How does bile help the pancreatic enzymes to work?
6. Where is food absorbed?
7. Which organ is the digested food taken to and in what blood vessel?
8. What happens to food that has not been digested?

ABSORPTION

This takes place through the **villi** of the small intestine.

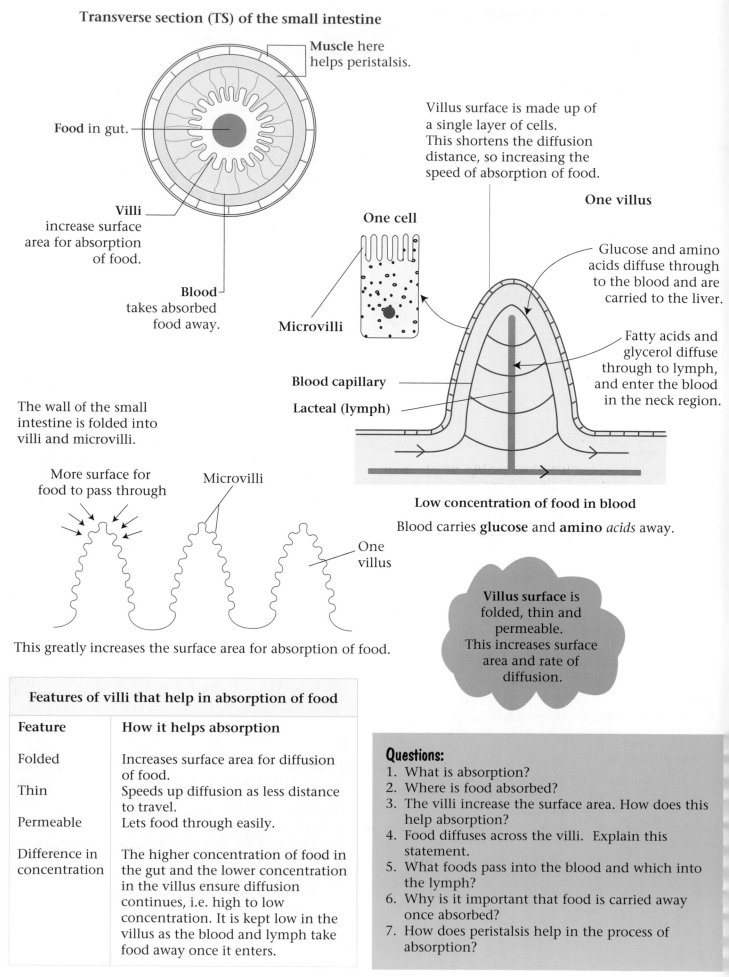

Transverse section (TS) of the small intestine

Muscle here helps peristalsis.

Food in gut.

Villi increase surface area for absorption of food.

Blood takes absorbed food away.

Villus surface is made up of a single layer of cells. This shortens the diffusion distance, so increasing the speed of absorption of food.

One villus

One cell

Microvilli

Glucose and amino acids diffuse through to the blood and are carried to the liver.

Fatty acids and glycerol diffuse through to lymph, and enter the blood in the neck region.

Blood capillary

Lacteal (lymph)

The wall of the small intestine is folded into villi and microvilli.

More surface for food to pass through

Microvilli

One villus

This greatly increases the surface area for absorption of food.

Low concentration of food in blood

Blood carries **glucose** and **amino** *acids* away.

Villus surface is folded, thin and permeable. This increases surface area and rate of diffusion.

Features of villi that help in absorption of food

Feature	How it helps absorption
Folded	Increases surface area for diffusion of food.
Thin	Speeds up diffusion as less distance to travel.
Permeable	Lets food through easily.
Difference in concentration	The higher concentration of food in the gut and the lower concentration in the villus ensure diffusion continues, i.e. high to low concentration. It is kept low in the villus as the blood and lymph take food away once it enters.

Questions:
1. What is absorption?
2. Where is food absorbed?
3. The villi increase the surface area. How does this help absorption?
4. Food diffuses across the villi. Explain this statement.
5. What foods pass into the blood and which into the lymph?
6. Why is it important that food is carried away once absorbed?
7. How does peristalsis help in the process of absorption?

STRUCTURE OF BLOOD

Blood

(Liquid part)

(Solid part) Cells

Plasma
A yellow liquid, 90% water, 10% dissolved substances (solutes).

- Carries food, i.e. glucose, amino acids.

- Carries enzymes, hormones and antibodies.

- Carries urea from liver to kidneys.

- Contains soluble fibrinogen involved in blood clotting.

Platelets (cell fragments) help the blood to clot when the skin is cut.

Fibrin threads

Platelet
Red blood cells get trapped in the threads. This forms a blood clot.
Blood clots stop:
- blood loss
- entry of germs.

Red blood cells carry oxygen

No nucleus

Red colour due to **haemoglobin** present.

Biconcave

White blood cells fight infections

In lungs
Oxygen enters red cells and joins with haemoglobin.

O_2 + Haemoglobin $\rightarrow$ Oxyhaemoglobin

Phagocytes

Lobed nucleus

Cell can change shape

These surround and digest bacteria.

Lymphocytes

Large nucleus

These produce antibodies which destroy bacteria and viruses.

In tissues
Oxygen leaves red cells to pass to respiring cells.

Oxyhaemoglobin $\rightarrow$ Haemoglobin + Oxygen

phagocyte

red blood cells

Human blood smear showing many red cells and the phagocyte with its lobed nucleus.

Metha: *Self-Assessment Colour Review of Clinical Haematology*

Carbon monoxide is emitted from car exhausts and from burning fossil fuels. It is extremely dangerous as it combines with haemoglobin permanently, so preventing the pick-up of oxygen.

Questions:
1. Which blood cells have no nucleus?
2. What is the job of the white blood cells?
3. Name three substances carried dissolved in the plasma.
4. Why is it important that our blood clots when we cut ourselves?
5. How do phagocytes kill bacteria?
6. Which blood cells carry oxygen?

BLOOD CELLS

Red blood cells (carry oxygen)

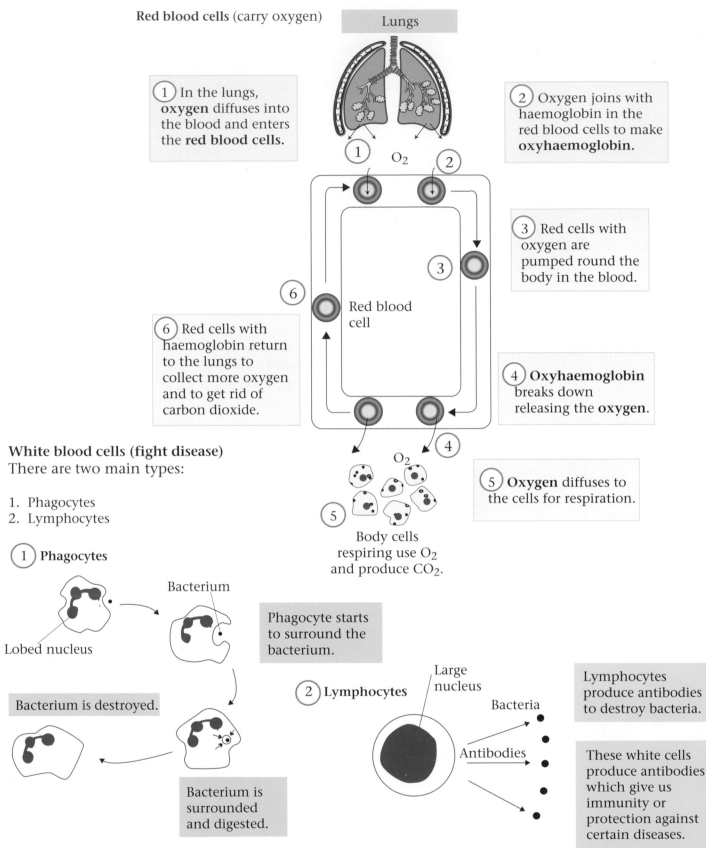

Lungs

① In the lungs, **oxygen** diffuses into the blood and enters the **red blood cells.**

② Oxygen joins with haemoglobin in the red blood cells to make **oxyhaemoglobin.**

O_2

③ Red cells with oxygen are pumped round the body in the blood.

Red blood cell

⑥ Red cells with haemoglobin return to the lungs to collect more oxygen and to get rid of carbon dioxide.

④ **Oxyhaemoglobin** breaks down releasing the **oxygen.**

O_2

⑤ **Oxygen** diffuses to the cells for respiration.

Body cells respiring use O_2 and produce CO_2.

White blood cells (fight disease)
There are two main types:

1. Phagocytes
2. Lymphocytes

① **Phagocytes**

Bacterium

Lobed nucleus

Phagocyte starts to surround the bacterium.

Bacterium is destroyed.

Bacterium is surrounded and digested.

② **Lymphocytes**

Large nucleus

Bacteria

Antibodies

Lymphocytes produce antibodies to destroy bacteria.

These white cells produce antibodies which give us immunity or protection against certain diseases.

Questions:
1. Where does oxygen pass into the blood?
2. Which substance does oxygen join with in the red blood cell and what is formed?
3. What pumps the red cells round the body?
4. How does oxygen reach respiring cells?
5. Which white cell has a lobed nucleus?
6. How do lymphocytes destroy bacteria?
7. Carbon monoxide, in cigarette smoke, can also join with haemoglobin. This joining is permanent. How will this affect a smoker?

MOVEMENT OF BLOOD AROUND THE BODY

Blood is carried in tubes called **blood vessels**.

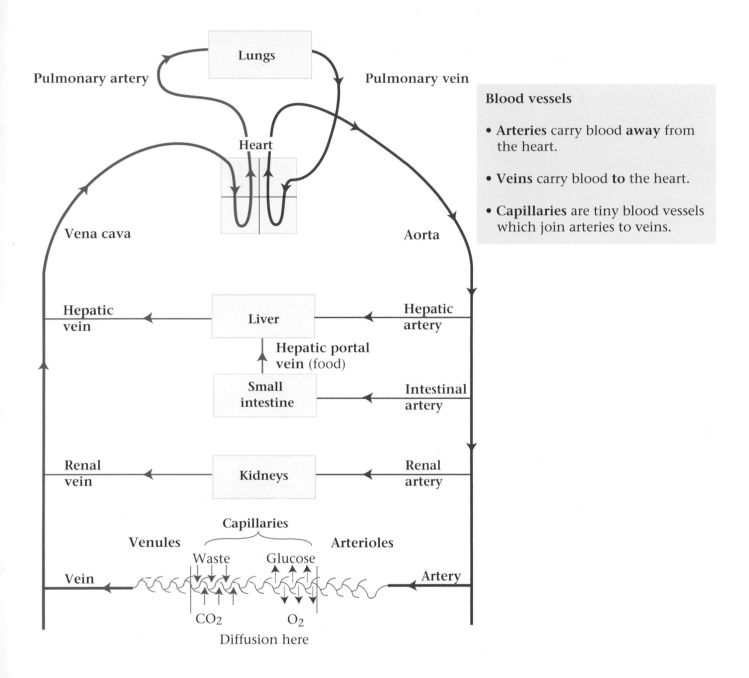

Blood vessels

- **Arteries** carry blood **away** from the heart.

- **Veins** carry blood **to** the heart.

- **Capillaries** are tiny blood vessels which join arteries to veins.

Capillaries are very thin, tiny blood vessels, with walls only one cell thick.

Here oxygen and glucose diffuse from the blood out to the cells. Carbon dioxide and waste, such as urea, diffuse from the body cells back into the blood. This is possible as capillaries are permeable.

Questions:
1. Which blood vessel enters the lungs?
2. What is added to blood in the lungs?
3. Which blood vessel carries oxygenated blood all round the body from the heart?
4. Why is the blood vessel leaving the kidneys deoxygenated?
5. What is carried in the hepatic portal vein?
6. Why do large arteries have to be split into tiny capillaries in body organs?
7. Which blood vessel leaves the liver, and what vessel does it join?

HEART (1)

Arteries carry blood **away** from the heart. Veins carry blood **to** the heart.

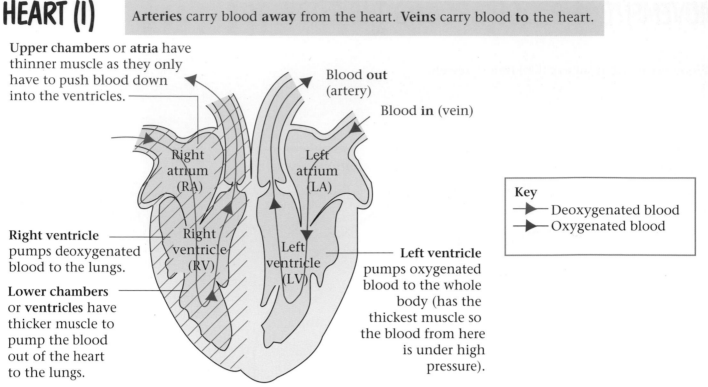

Upper chambers or **atria** have thinner muscle as they only have to push blood down into the ventricles.

Blood **out** (artery)

Blood **in** (vein)

Right atrium (RA)

Left atrium (LA)

Right ventricle pumps deoxygenated blood to the lungs.

Right ventricle (RV)

Left ventricle (LV)

Lower chambers or **ventricles** have thicker muscle to pump the blood out of the heart to the lungs.

Left ventricle pumps oxygenated blood to the whole body (has the thickest muscle so the blood from here is under high pressure).

Key
→ Deoxygenated blood
▶ Oxygenated blood

The heart is made of **cardiac muscle**. Beating starts in the wall of the right atrium at the **pacemaker**. The human heart beats about 70 times in a minute and causes the pulse which can be felt in the wrist.

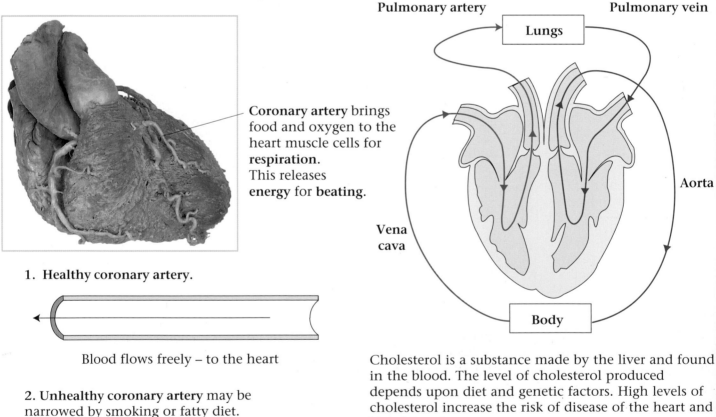

Pulmonary artery

Lungs

Pulmonary vein

Aorta

Vena cava

Body

Courtesy of Ralph Hutchings.

Coronary artery brings food and oxygen to the heart muscle cells for **respiration**. This releases **energy** for **beating**.

1. **Healthy coronary artery.**

Blood flows freely – to the heart

2. **Unhealthy coronary artery** may be narrowed by smoking or fatty diet.

Blood clot ←

Artery narrowed

Blood clot can cause a blockage if the artery is narrow. If blood does not reach the heart, the muscle stops beating. This causes a **heart attack**.

Cholesterol is a substance made by the liver and found in the blood. The level of cholesterol produced depends upon diet and genetic factors. High levels of cholesterol increase the risk of disease of the heart and blood vessels.

Questions:
1. Which chamber of the heart has the thickest muscle and why?
2. What type of muscle makes up the heart?
3. Which blood vessel supplies the heart muscle itself with food and oxygen?

86

HEART (II)

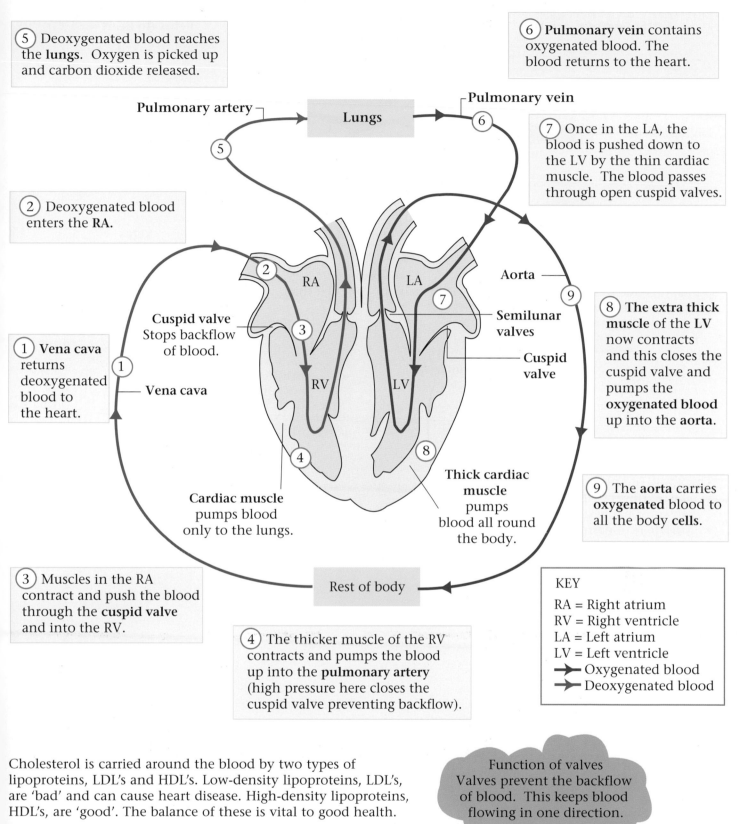

⑤ Deoxygenated blood reaches the **lungs**. Oxygen is picked up and carbon dioxide released.

⑥ **Pulmonary vein** contains oxygenated blood. The blood returns to the heart.

Pulmonary artery — Lungs — Pulmonary vein

⑦ Once in the LA, the blood is pushed down to the LV by the thin cardiac muscle. The blood passes through open cuspid valves.

② Deoxygenated blood enters the **RA**.

Aorta

① **Vena cava** returns deoxygenated blood to the heart.

Cuspid valve Stops backflow of blood.

Vena cava

Semilunar valves

Cuspid valve

⑧ **The extra thick muscle** of the LV now contracts and this closes the cuspid valve and pumps the **oxygenated blood** up into the **aorta**.

Cardiac muscle pumps blood only to the lungs.

Thick cardiac muscle pumps blood all round the body.

⑨ The **aorta** carries **oxygenated** blood to all the body **cells**.

③ Muscles in the RA contract and push the blood through the **cuspid valve** and into the RV.

Rest of body

KEY
RA = Right atrium
RV = Right ventricle
LA = Left atrium
LV = Left ventricle
→ Oxygenated blood
→ Deoxygenated blood

④ The thicker muscle of the RV contracts and pumps the blood up into the **pulmonary artery** (high pressure here closes the cuspid valve preventing backflow).

Cholesterol is carried around the blood by two types of lipoproteins, LDL's and HDL's. Low-density lipoproteins, LDL's, are 'bad' and can cause heart disease. High-density lipoproteins, HDL's, are 'good'. The balance of these is vital to good health.

Function of valves
Valves prevent the backflow of blood. This keeps blood flowing in one direction.

The heart is a double pump, a **right** and **left** pump

Side of heart	Pressure	Blood	Sent to
Right Ventricle	Lower	Deoxygenated	Lungs
Left Ventricle	Higher	Oxygenated	Body

Questions:
1. What causes the higher pressure on the left side of the heart?
2. What happens to blood when it reaches the lungs?
3. What is the job of the valves in the heart?

BLOOD VESSELS

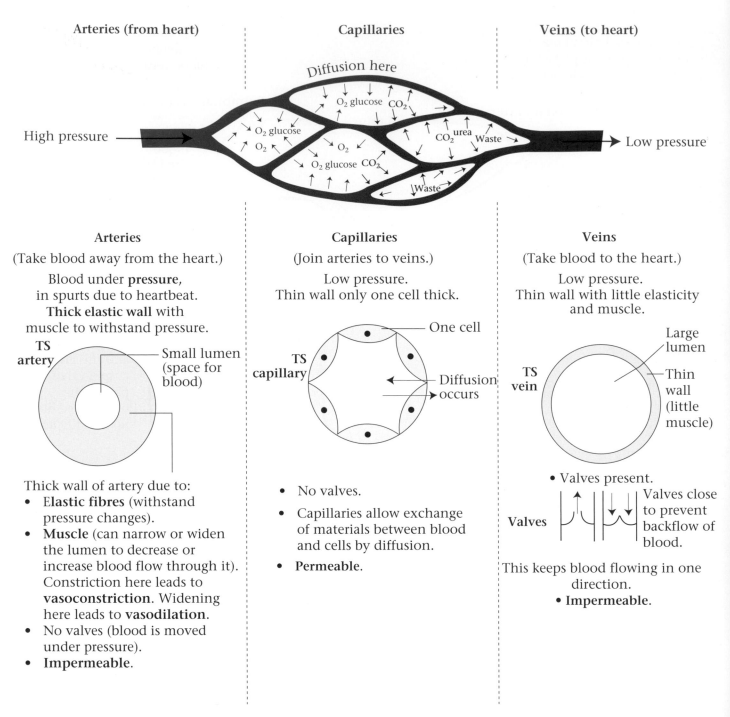

Arteries (from heart) — **Capillaries** — **Veins (to heart)**

Diffusion here

High pressure → — Low pressure

O₂ glucose CO₂
O₂ glucose
O₂
O₂ glucose CO₂
CO₂ urea Waste
Waste

Arteries
(Take blood away from the heart.)

Blood under **pressure**, in spurts due to heartbeat. **Thick elastic wall** with muscle to withstand pressure.

TS artery — Small lumen (space for blood)

Thick wall of artery due to:
- **Elastic fibres** (withstand pressure changes).
- **Muscle** (can narrow or widen the lumen to decrease or increase blood flow through it). Constriction here leads to **vasoconstriction**. Widening here leads to **vasodilation**.
- No valves (blood is moved under pressure).
- **Impermeable**.

Capillaries
(Join arteries to veins.)

Low pressure.
Thin wall only one cell thick.

TS capillary — One cell
— Diffusion occurs

- No valves.
- Capillaries allow exchange of materials between blood and cells by diffusion.
- **Permeable**.

Veins
(Take blood to the heart.)

Low pressure.
Thin wall with little elasticity and muscle.

TS vein — Large lumen
— Thin wall (little muscle)

- Valves present.

Valves — Valves close to prevent backflow of blood.

This keeps blood flowing in one direction.
- **Impermeable**.

Oxygen and **glucose** diffuse from the blood to the body cells. **Carbon dioxide** and **urea** diffuse from the cells back into the blood. This exchange of material takes place through the permeable capillaries.

Questions:
1. Which blood vessels take blood to the heart?
2. How do the valves work in veins, and why are they necessary?
3. Which blood vessels are permeable?
4. Name two substances which might diffuse out of blood and into the body cells.
5. What causes the high pressure in arteries?
6. Why must arteries have thick, elastic walls?

CAPILLARIES Exchange of materials.

Capillaries – tiny blood vessels.

Thin
Permeable   } Allow diffusion of materials
Large surface area   between blood and cells.

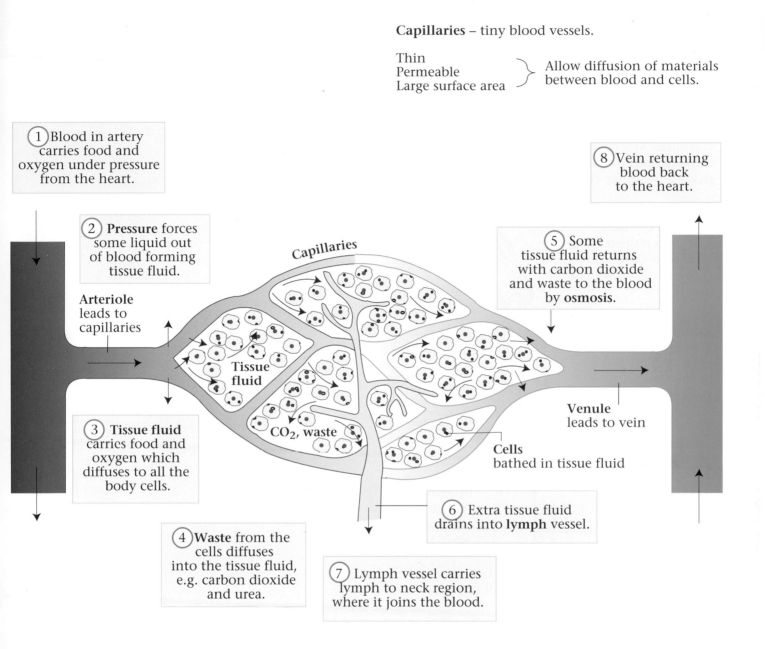

1 Blood in artery carries food and oxygen under pressure from the heart.

2 **Pressure** forces some liquid out of blood forming tissue fluid.

Arteriole leads to capillaries

Capillaries

Tissue fluid

CO_2, waste

3 **Tissue fluid** carries food and oxygen which diffuses to all the body cells.

4 **Waste** from the cells diffuses into the tissue fluid, e.g. carbon dioxide and urea.

5 Some tissue fluid returns with carbon dioxide and waste to the blood by **osmosis**.

8 Vein returning blood back to the heart.

Venule leads to vein

Cells bathed in tissue fluid

6 Extra tissue fluid drains into **lymph** vessel.

7 Lymph vessel carries lymph to neck region, where it joins the blood.

- Blood in the artery is under pressure.
- The artery divides into smaller arterioles and capillaries.
- Some of the liquid part of the blood, plasma is pushed through the permeable capillary wall to form **tissue fluid**. (Blood cells and plasma proteins are too large to filter through.)
- Oxygen and glucose in the tissue fluid diffuse to the body cells.
- Carbon dioxide and urea diffuse from the body cells into the tissue fluid.
- At the venule end of the capillaries, some tissue fluid returns to the blood by **osmosis**.
- The remaining tissue fluid drains into lymph vessels to form **lymph**.

Questions:
1. What is tissue fluid and how is it formed?
2. How do body cells get their oxygen and food?
3. Why must capillaries be permeable?
4. Body cells produce carbon dioxide and waste, how does this return to the blood?
5. What happens to the tissue fluid that does not return to the blood immediately?

GAS EXCHANGE AND RESPIRATION

THE HUMAN THORAX

Pleural membranes
Contain a fluid which cushions the lungs.

Trachea (windpipe)
This carries air down to the lungs. Mucus here traps dirt and bacteria, helping to keep the lungs clean.

Rings of cartilage
These keep the trachea open and allow movement of the neck.

Ribs (cut)
These protect the heart and lungs in the thorax from damage.

Intercostal muscles (between the ribs)
These move the rib cage up and down during breathing.

Pleural fluid
Allows lungs to move freely.

Larynx (voice box)
Infection here causes laryngitis.

In the nose
1. Air is cleaned by cilia hairs.
2. Air is warmed by blood vessels.
3. Air is moistened by mucus.

Bronchus
This tube enters the lung. Infection here causes bronchitis. It is supported by rings of cartilage.

Left lung

Heart
Pumps blood round the body.

Bronchioles
Tiny tubes in the lungs carrying air down to the alveoli.

Alveoli (air sacs)
These allow gas exchange between the lungs and the blood. Each is surrounded by a network of blood capillaries taking oxygen away.

Diaphragm
This muscle moves up and down causing air to rush into or out of the lungs (**ventilation**).

From McMinn et al: Concise Handbook of Human Anatomy

X-ray of human thorax clearly showing the ribs and lungs (dark shadow). The position of the heart can be seen by the displaced lungs on the right side.

Questions:
1. What protects the lungs and heart?
2. What is the role of mucus in the trachea?
3. What happens to air in the nose?
4. Describe the function of the alveoli.
5. What is the job of the rings of cartilage?

GAS EXCHANGE IN THE ALVEOLI

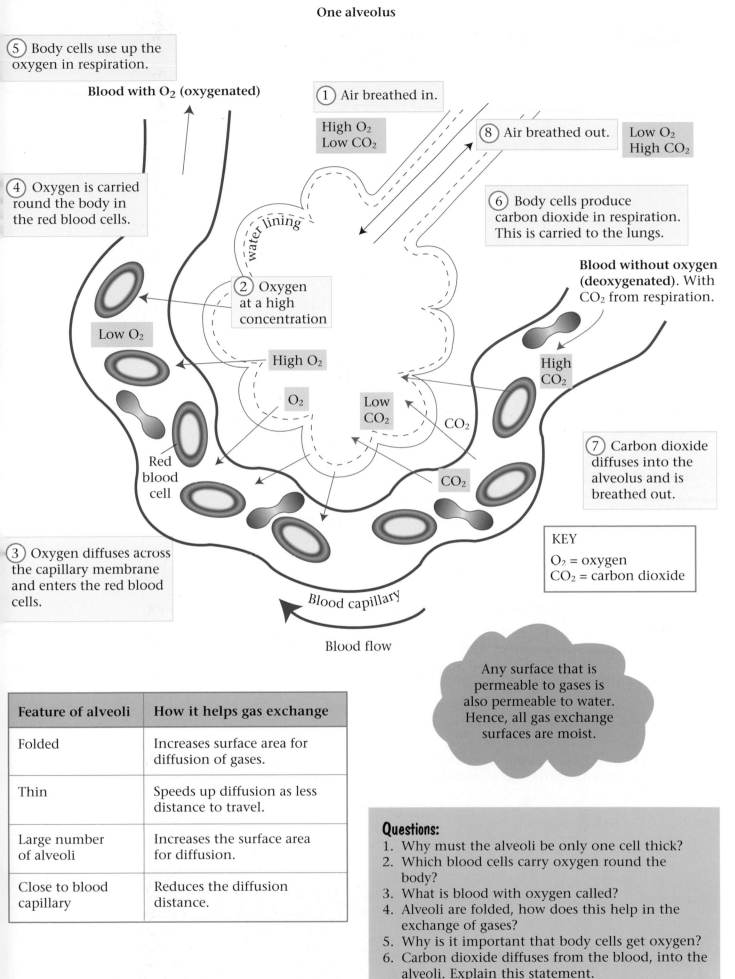

One alveolus

(5) Body cells use up the oxygen in respiration.

Blood with O₂ (oxygenated)

(1) Air breathed in.

High O₂
Low CO₂

(8) Air breathed out.

Low O₂
High CO₂

(4) Oxygen is carried round the body in the red blood cells.

water lining

(6) Body cells produce carbon dioxide in respiration. This is carried to the lungs.

Blood without oxygen (deoxygenated). With CO₂ from respiration.

(2) Oxygen at a high concentration

Low O₂

High O₂

High CO₂

O₂

Low CO₂

CO₂

Red blood cell

CO₂

(7) Carbon dioxide diffuses into the alveolus and is breathed out.

KEY
O₂ = oxygen
CO₂ = carbon dioxide

(3) Oxygen diffuses across the capillary membrane and enters the red blood cells.

Blood capillary

Blood flow

Any surface that is permeable to gases is also permeable to water. Hence, all gas exchange surfaces are moist.

Feature of alveoli	How it helps gas exchange
Folded	Increases surface area for diffusion of gases.
Thin	Speeds up diffusion as less distance to travel.
Large number of alveoli	Increases the surface area for diffusion.
Close to blood capillary	Reduces the diffusion distance.

Questions:
1. Why must the alveoli be only one cell thick?
2. Which blood cells carry oxygen round the body?
3. What is blood with oxygen called?
4. Alveoli are folded, how does this help in the exchange of gases?
5. Why is it important that body cells get oxygen?
6. Carbon dioxide diffuses from the blood, into the alveoli. Explain this statement.

BREATHING (I)

Changes in the position of the ribs and diaphragm when breathing.

Breathing out
Breathing in
Position when breathing out.

Rib cage
Diaphragm

Position when breathing in

Increase in volume caused by movement of diaphragm and rib cage.

Gas	Air breathed in	Air breathed out	Changes
Nitrogen	79%	79%	None.
Oxygen	20%	16%	Oxygen enters blood so less is breathed out.
Carbon dioxide	0.04%	4%	Carbon dioxide is produced by respiring cells – so more carbon dioxide is breathed out.
Water vapour	Varies	Saturated	Water is produced by moist cells lining lungs and in respiration.

Air always moves from high to low pressure.

Breathing in:

Diaphragm contracts and flattens

↓

Intercostal muscles pull ribs up and out

↓

Both increase volume of thorax

↓

This lowers the pressure in the thorax – air outside has higher pressure

↓

Air rushes into lungs

Breathing out:

Diaphragm relaxes and rises to a dome shape

↓

Intercostal muscles pull ribs down and in

↓

Both decrease volume of thorax

↓

This increases the pressure in the thorax – air outside has lower pressure.

↓

Air is forced out of lungs

Questions:
1. What two actions increase the volume of the thorax?
2. How does an increase in volume affect the pressure?
3. What causes air to rush into the lungs?
4. Which gas is taken into the body during breathing?
5. Give three differences between air breathed in and air breathed out?
6. Why is air forced out of the body when the diaphragm rises into a dome shape?

BREATHING (II)

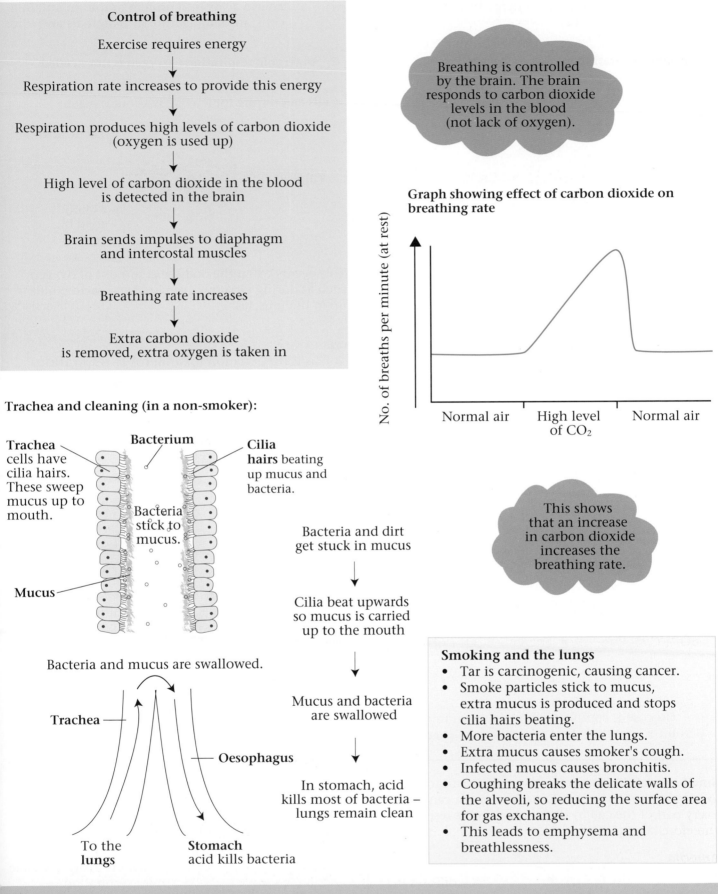

Control of breathing

Exercise requires energy

↓

Respiration rate increases to provide this energy

↓

Respiration produces high levels of carbon dioxide
(oxygen is used up)

↓

High level of carbon dioxide in the blood
is detected in the brain

↓

Brain sends impulses to diaphragm
and intercostal muscles

↓

Breathing rate increases

↓

Extra carbon dioxide
is removed, extra oxygen is taken in

Breathing is controlled
by the brain. The brain
responds to carbon dioxide
levels in the blood
(not lack of oxygen).

**Graph showing effect of carbon dioxide on
breathing rate**

No. of breaths per minute (at rest)

Normal air | High level of CO_2 | Normal air

This shows
that an increase
in carbon dioxide
increases the
breathing rate.

Trachea and cleaning (in a non-smoker):

Trachea
cells have
cilia hairs.
These sweep
mucus up to
mouth.

Bacterium

**Cilia
hairs** beating
up mucus and
bacteria.

Bacteria
stick to
mucus.

Mucus

Bacteria and dirt
get stuck in mucus

↓

Cilia beat upwards
so mucus is carried
up to the mouth

↓

Mucus and bacteria
are swallowed

↓

In stomach, acid
kills most of bacteria –
lungs remain clean

Bacteria and mucus are swallowed.

Trachea

Oesophagus

To the
lungs

Stomach
acid kills bacteria

Smoking and the lungs
- Tar is carcinogenic, causing cancer.
- Smoke particles stick to mucus,
  extra mucus is produced and stops
  cilia hairs beating.
- More bacteria enter the lungs.
- Extra mucus causes smoker's cough.
- Infected mucus causes bronchitis.
- Coughing breaks the delicate walls of
  the alveoli, so reducing the surface area
  for gas exchange.
- This leads to emphysema and
  breathlessness.

Questions:
1. What controls the breathing rate?
2. What increases the breathing rate?
3. How do trachea cells clean the air breathed in?
4. Why is it better to swallow bacteria than breathe
   them in?
5. Which part of cigarette smoke causes cancer?
6. Why do smokers suffer from more lung
   infections?

RESPIRATION Aerobic and anaerobic.

All cells respire to release the energy from food (usually sugar).

The energy released during respiration is used for muscle contraction, to build up amino acids into protein and to maintain a constant temperature in mammals and birds.

1. Aerobic respiration – oxygen is needed.

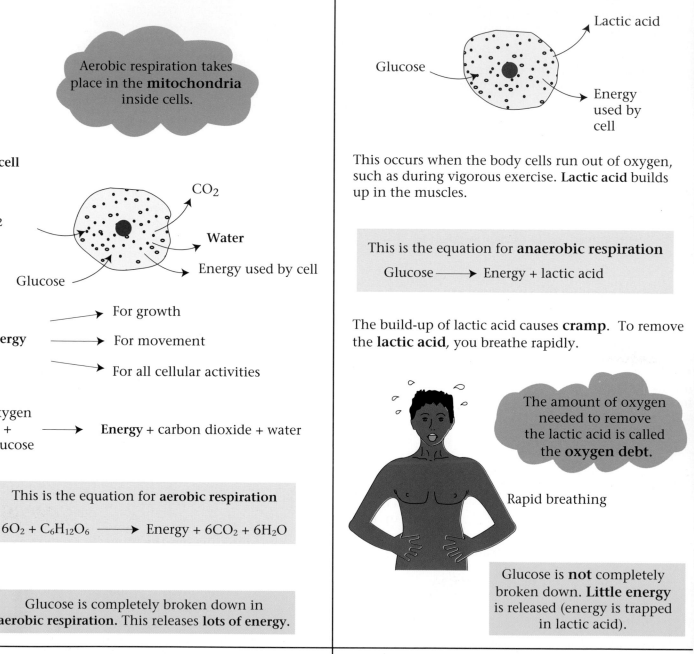

Aerobic respiration takes place in the **mitochondria** inside cells.

A cell

O_2

CO_2

Water

Energy used by cell

Glucose

Energy → For growth

Energy → For movement

Energy → For all cellular activities

Oxygen + Glucose → **Energy + carbon dioxide + water**

This is the equation for **aerobic respiration**

$$6O_2 + C_6H_{12}O_6 \longrightarrow Energy + 6CO_2 + 6H_2O$$

Glucose is completely broken down in **aerobic respiration**. This releases **lots of energy**.

2. Anaerobic respiration – oxygen is **not** needed

Cells can respire for a while without oxygen.

Lactic acid

Glucose

Energy used by cell

This occurs when the body cells run out of oxygen, such as during vigorous exercise. **Lactic acid** builds up in the muscles.

This is the equation for **anaerobic respiration**

Glucose ⟶ Energy + lactic acid

The build-up of lactic acid causes **cramp**. To remove the **lactic acid**, you breathe rapidly.

The amount of oxygen needed to remove the lactic acid is called the **oxygen debt**.

Rapid breathing

Glucose is **not** completely broken down. **Little energy** is released (energy is trapped in lactic acid).

Small animals like *Amoeba* have a **large surface area** compared to their volume, so oxygen can reach every part of the cell by **diffusion**. Aerobic respiration therefore takes place.

Amoeba

O_2 diffuses in

Yeast can respire anaerobically and releases alcohol. This is used to make wine.

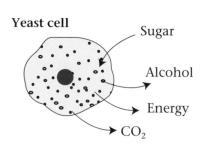

Yeast cell

Sugar

Alcohol

Energy

CO_2

Yeast lives on the skin of grapes where sugar is easily available for respiration. The carbon dioxide is used in baking to make bread rise.

94

DRUGS Drugs are substances that alter the way the body works.

Depressants (slow down transmission of nerves at the synapse)

1. Alcohol

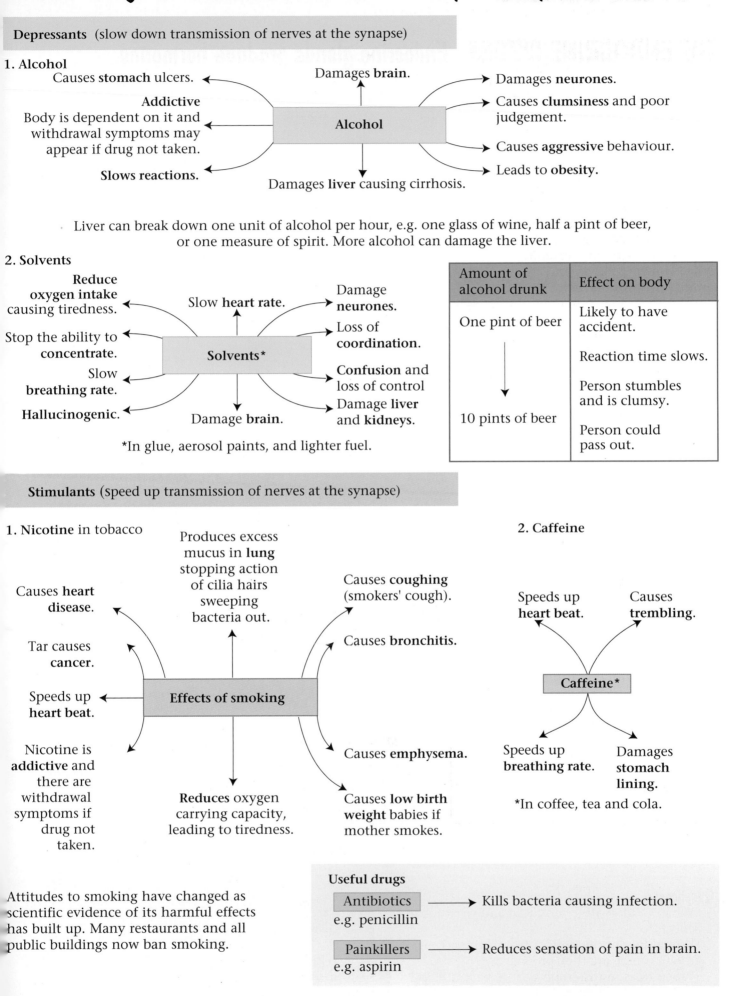

Causes **stomach** ulcers.

Addictive
Body is dependent on it and withdrawal symptoms may appear if drug not taken.

Slows reactions.

Damages **brain.**

Alcohol

Damages **neurones.**

Causes **clumsiness** and poor judgement.

Causes **aggressive** behaviour.

Leads to **obesity.**

Damages **liver** causing cirrhosis.

Liver can break down one unit of alcohol per hour, e.g. one glass of wine, half a pint of beer, or one measure of spirit. More alcohol can damage the liver.

2. Solvents

Reduce oxygen intake causing tiredness.

Stop the ability to **concentrate.**

Slow **breathing rate.**

Hallucinogenic.

Slow **heart rate.**

Solvents*

Damage **brain.**

Damage **neurones.**

Loss of **coordination.**

Confusion and loss of control

Damage **liver** and **kidneys.**

*In glue, aerosol paints, and lighter fuel.

Amount of alcohol drunk	Effect on body
One pint of beer ↓ 10 pints of beer	Likely to have accident. Reaction time slows. Person stumbles and is clumsy. Person could pass out.

Stimulants (speed up transmission of nerves at the synapse)

1. Nicotine in tobacco

Causes **heart disease.**

Tar causes **cancer.**

Speeds up **heart beat.**

Nicotine is **addictive** and there are withdrawal symptoms if drug not taken.

Produces excess mucus in **lung** stopping action of cilia hairs sweeping bacteria out.

Effects of smoking

Reduces oxygen carrying capacity, leading to tiredness.

Causes **coughing** (smokers' cough).

Causes **bronchitis.**

Causes **emphysema.**

Causes **low birth weight** babies if mother smokes.

2. Caffeine

Speeds up **heart beat.**

Causes **trembling.**

Caffeine*

Speeds up **breathing rate.**

Damages stomach lining.

*In coffee, tea and cola.

Attitudes to smoking have changed as scientific evidence of its harmful effects has built up. Many restaurants and all public buildings now ban smoking.

Useful drugs

Antibiotics
e.g. penicillin — Kills bacteria causing infection.

Painkillers
e.g. aspirin — Reduces sensation of pain in brain.

95

HOMEOSTASIS

THE ENDOCRINE SYSTEM   Endocrine glands produce hormones.

Endocrine gland → Hormone (a chemical messenger) → Blood → Target organs

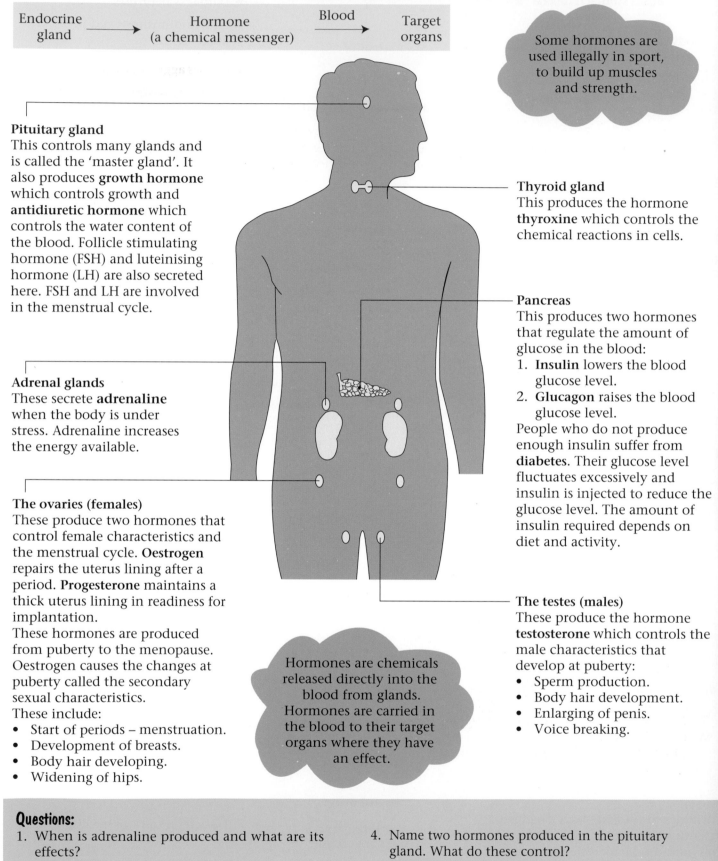

Some hormones are used illegally in sport, to build up muscles and strength.

Pituitary gland
This controls many glands and is called the 'master gland'. It also produces **growth hormone** which controls growth and **antidiuretic hormone** which controls the water content of the blood. Follicle stimulating hormone (FSH) and luteinising hormone (LH) are also secreted here. FSH and LH are involved in the menstrual cycle.

Thyroid gland
This produces the hormone **thyroxine** which controls the chemical reactions in cells.

Pancreas
This produces two hormones that regulate the amount of glucose in the blood:
1. **Insulin** lowers the blood glucose level.
2. **Glucagon** raises the blood glucose level.
People who do not produce enough insulin suffer from **diabetes**. Their glucose level fluctuates excessively and insulin is injected to reduce the glucose level. The amount of insulin required depends on diet and activity.

Adrenal glands
These secrete **adrenaline** when the body is under stress. Adrenaline increases the energy available.

The ovaries (females)
These produce two hormones that control female characteristics and the menstrual cycle. **Oestrogen** repairs the uterus lining after a period. **Progesterone** maintains a thick uterus lining in readiness for implantation.
These hormones are produced from puberty to the menopause. Oestrogen causes the changes at puberty called the secondary sexual characteristics.
These include:
• Start of periods – menstruation.
• Development of breasts.
• Body hair developing.
• Widening of hips.

Hormones are chemicals released directly into the blood from glands. Hormones are carried in the blood to their target organs where they have an effect.

The testes (males)
These produce the hormone **testosterone** which controls the male characteristics that develop at puberty:
• Sperm production.
• Body hair development.
• Enlarging of penis.
• Voice breaking.

Questions:
1. When is adrenaline produced and what are its effects?
2. Which two hormones control the blood sugar level? Where are they produced?
3. When does testosterone start to be produced?
4. Name two hormones produced in the pituitary gland. What do these control?
5. How do hormones reach their target organs?
6. Which two hormones are produced in the ovaries?

ADRENAL GLANDS

These produce the hormone **adrenaline**. Adrenaline prepares the body for emergencies, e.g. fighting or running away, by making more energy available. (It is known as the 'flight or fight' hormone.)

Adrenaline is released whenever the body is under stress, e.g. when we are angry, nervous, or terrified. Adrenaline diffuses from the adrenal glands into the blood and is carried to several target organs where it has its effects.

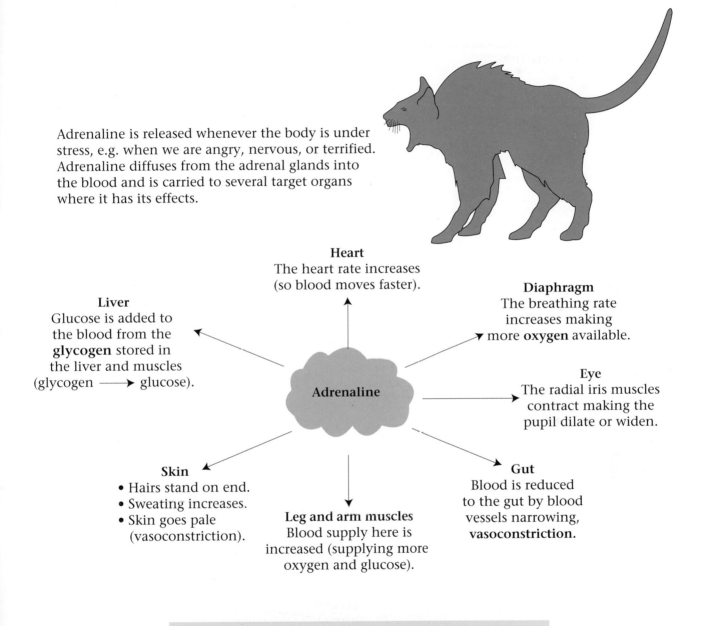

Heart
The heart rate increases (so blood moves faster).

Liver
Glucose is added to the blood from the **glycogen** stored in the liver and muscles (glycogen ⟶ glucose).

Diaphragm
The breathing rate increases making more **oxygen** available.

Adrenaline

Eye
The radial iris muscles contract making the pupil dilate or widen.

Skin
• Hairs stand on end.
• Sweating increases.
• Skin goes pale (vasoconstriction).

Leg and arm muscles
Blood supply here is increased (supplying more oxygen and glucose).

Gut
Blood is reduced to the gut by blood vessels narrowing, **vasoconstriction**.

The blood contains more oxygen (from increased breathing) and more glucose (from the liver). Therefore more glucose and oxygen reaches the muscles enabling the respiration rate there to increase. Increased respiration results in more energy available.

Questions:
1. Before a race athletes are nervous. Why is this an advantage for them?
2. Name four target organs affected by adrenaline.
3. How is the blood in the gut decreased?

CONTROL OF BLOOD SUGAR LEVEL
An example of homeostasis keeping a constant internal environment.

The sugar carried in the blood is glucose.

Digestion and absorption of carbohydrates **adds** glucose to the blood.

During exercise lots of glucose is **removed** from blood by muscle cells to release energy in respiration.

1. The role of insulin

2. The role of glucagon

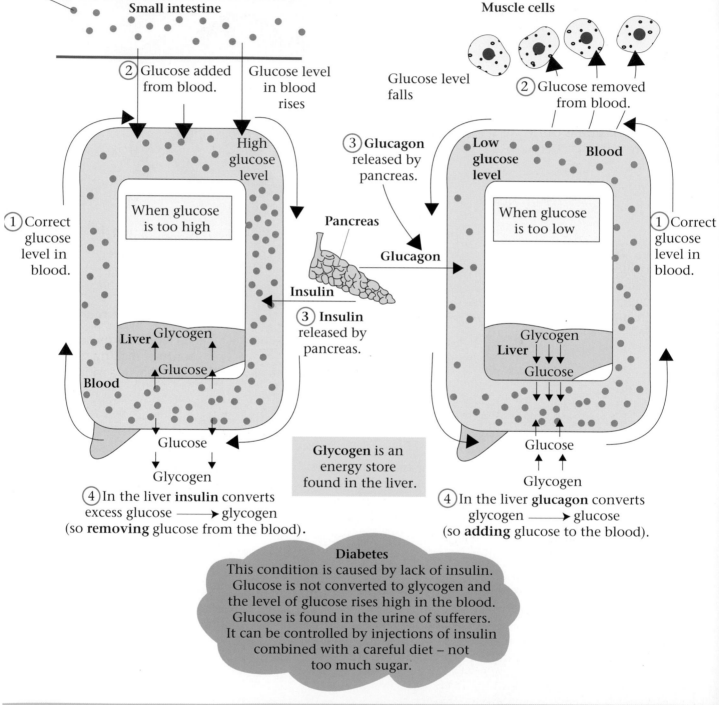

Glucose

Small intestine

② Glucose added from blood.

Glucose level in blood rises

Muscle cells

Glucose level falls

② Glucose removed from blood.

③ **Glucagon** released by pancreas.

Low glucose level

Blood

High glucose level

① Correct glucose level in blood.

When glucose is too high

Pancreas

Glucagon

① Correct glucose level in blood.

When glucose is too low

Insulin

③ **Insulin** released by pancreas.

Liver Glycogen

Glucose

Blood

Glycogen

Liver

Glucose

Glucose

Glycogen

Glycogen is an energy store found in the liver.

④ In the liver **insulin** converts excess glucose ——▶ glycogen (so **removing** glucose from the blood).

Glucose

Glycogen

④ In the liver **glucagon** converts glycogen ——▶ glucose (so **adding** glucose to the blood).

Diabetes
This condition is caused by lack of insulin.
Glucose is not converted to glycogen and the level of glucose rises high in the blood.
Glucose is found in the urine of sufferers.
It can be controlled by injections of insulin combined with a careful diet – not too much sugar.

Questions:
1. Which hormone reduces the glucose level and how does it do this?
2. Where is glucose added to the blood?
3. How is glucose used up in the body?
4. Which hormone increases the glucose level in the blood?
5. Where are the hormones produced that control glucose levels?
6. Excess glucose is stored as glycogen in which organ?

HOMEOSTASIS AND THE LIVER

The liver keeps the level of amino acids and glucose at a constant level, i.e. homeostasis.

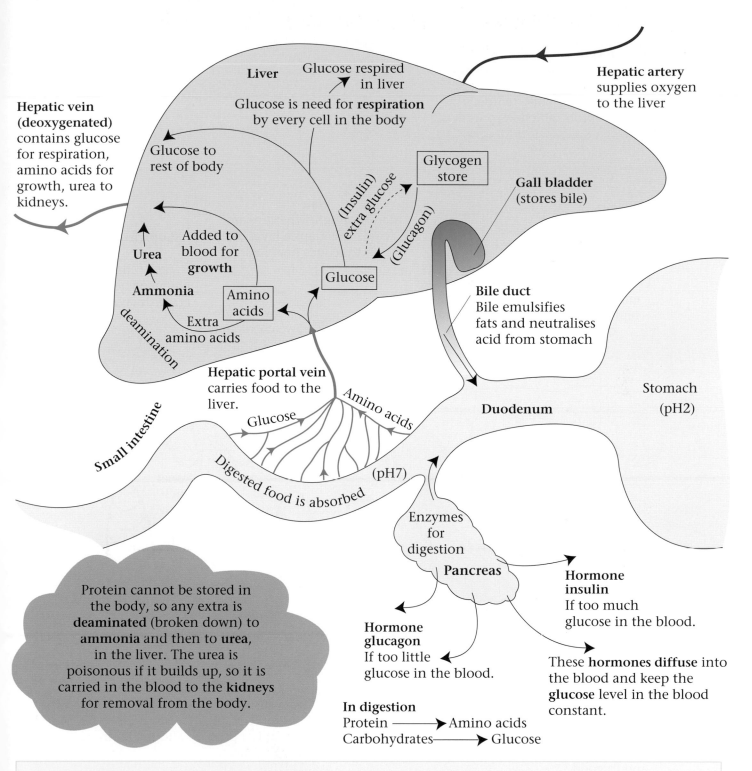

Hepatic vein (deoxygenated) contains glucose for respiration, amino acids for growth, urea to kidneys.

Liver

Glucose respired in liver

Glucose is need for **respiration** by every cell in the body

Glucose to rest of body

(Insulin) extra glucose

Glycogen store

(Glucagon)

Glucose

Gall bladder (stores bile)

Hepatic artery supplies oxygen to the liver

Urea

Added to blood for **growth**

Ammonia

deamination

Extra amino acids

Amino acids

Bile duct Bile emulsifies fats and neutralises acid from stomach

Hepatic portal vein carries food to the liver.

Amino acids

Glucose

Duodenum

Stomach (pH2)

Small intestine

Digested food is absorbed (pH7)

Enzymes for digestion

Pancreas

Hormone insulin If too much glucose in the blood.

Hormone glucagon If too little glucose in the blood.

These **hormones diffuse** into the blood and keep the **glucose** level in the blood constant.

Protein cannot be stored in the body, so any extra is **deaminated** (broken down) to **ammonia** and then to **urea**, in the liver. The urea is poisonous if it builds up, so it is carried in the blood to the **kidneys** for removal from the body.

In digestion
Protein ⟶ Amino acids
Carbohydrates ⟶ Glucose

Glucose and amino acids are carried in the hepatic portal vein to the liver.

Fate of glucose
- Some glucose is respired by the liver.
- Some glucose is added to the hepatic vein and carried to other body cells, for respiration.
- Excess glucose is converted to **glycogen** in the liver and stored.
- The hormone **insulin**, secreted by the pancreas, converts glucose to glycogen.
- If glucose levels fall in the blood, the hormone **glucagon** from the pancreas, causes the conversion of glycogen back to glucose.

Fate of amino acids
- Some amino acids are used by the liver cells for growth.
- Some amino acids are added to the hepatic vein and carried to other body cells for growth.
- Excess amino acids are deaminated, broken down to ammonia and then converted to urea which is less poisonous (in the liver).
- Urea is added to the hepatic vein and carried to the kidneys for removal in the urine.

THE SKIN

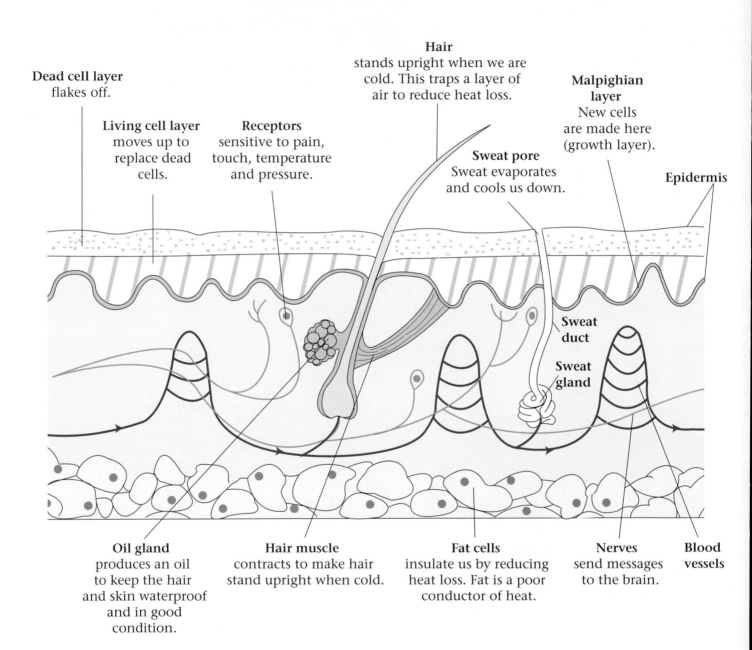

Hair
stands upright when we are cold. This traps a layer of air to reduce heat loss.

Dead cell layer
flakes off.

Living cell layer
moves up to replace dead cells.

Receptors
sensitive to pain, touch, temperature and pressure.

Malpighian layer
New cells are made here (growth layer).

Sweat pore
Sweat evaporates and cools us down.

Epidermis

Sweat duct

Sweat gland

Oil gland
produces an oil to keep the hair and skin waterproof and in good condition.

Hair muscle
contracts to make hair stand upright when cold.

Fat cells
insulate us by reducing heat loss. Fat is a poor conductor of heat.

Nerves
send messages to the brain.

Blood vessels

Excess fat from food is stored as fat under the skin.

Questions:
1. Where are new cells made in the skin?
2. Where is oil produced in the skin and why?
3. How does the hair and fat in the skin help to keep us warm?
4. What is produced when we are too hot, to cool us down?
5. Give two functions of the skin.
6. What is the skin sensitive to?

TEMPERATURE REGULATION This keeps the body temperature constant.

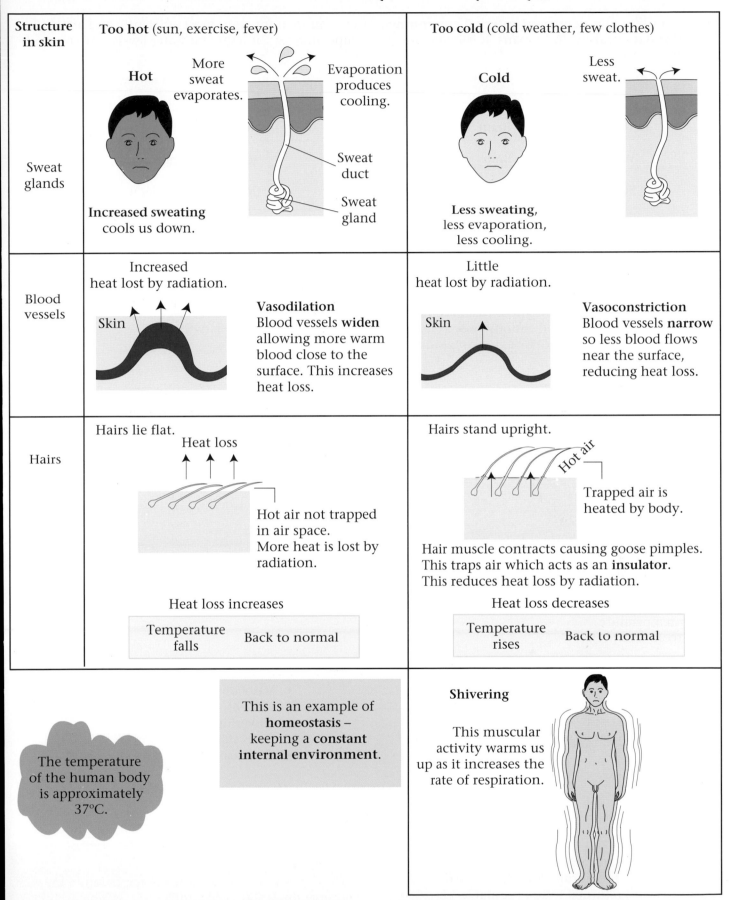

Structure in skin	Too hot (sun, exercise, fever)	Too cold (cold weather, few clothes)
Sweat glands	Hot · More sweat evaporates. · Evaporation produces cooling. · Sweat duct · Sweat gland · **Increased sweating** cools us down.	Cold · Less sweat. · **Less sweating,** less evaporation, less cooling.
Blood vessels	Increased heat lost by radiation. · Skin · **Vasodilation** Blood vessels **widen** allowing more warm blood close to the surface. This increases heat loss.	Little heat lost by radiation. · Skin · **Vasoconstriction** Blood vessels **narrow** so less blood flows near the surface, reducing heat loss.
Hairs	Hairs lie flat. · Heat loss · Hot air not trapped in air space. More heat is lost by radiation. · Heat loss increases · Temperature falls — Back to normal	Hairs stand upright. · Hot air · Trapped air is heated by body. · Hair muscle contracts causing goose pimples. This traps air which acts as an **insulator.** This reduces heat loss by radiation. · Heat loss decreases · Temperature rises — Back to normal

This is an example of **homeostasis –** keeping a **constant internal environment.**

The temperature of the human body is approximately 37°C.

Shivering

This muscular activity warms us up as it increases the rate of respiration.

Questions:
1. What makes us too hot?
2. What is vasodilation and how does it increase heat loss?
3. When does the hair muscle contract and what does this achieve?
4. How does vasoconstriction reduce heat loss?

HOMEOSTASIS e.g. control of body temperature.

This means keeping a **constant internal environment**. Cells can then function most efficiently and an organism stays healthy. In humans, it is important that temperature, sugar level and water levels stay constant.

Condition	Controller	Organ involved	Hormone
Temperature	Brain	Skin	None
Water level	Brain	Kidneys	Anti-diuretic hormone (ADH)
Sugar level	Pancreas	Liver	Insulin and glucagon

Changes in the above conditions are detected and brought back to their optimum level.

It is important that temperature remains constant for enzyme action in cells. Mammals and birds must maintain a constant temperature to survive. They are called **endothermic** as they control their temperature from within the body, not by external means.

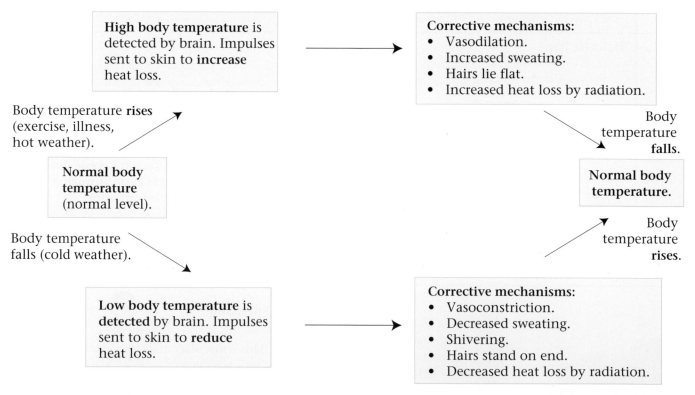

This is maintained by **negative feedback mechanisms**.

Definition of a negative feedback mechanism.
Any change in the normal level is detected, corrected, and returned to the normal level.

Questions:
1. Why is it important that the temperature in mammals stays constant?
2. Which two groups of vertebrates have a constant body temperature?
3. What controls the temperature of the body?
4. What is vasoconstriction and when does it occur?

EXCRETION IN HUMANS
Excretion is the removal of waste produced by cells.

Excretory wastes

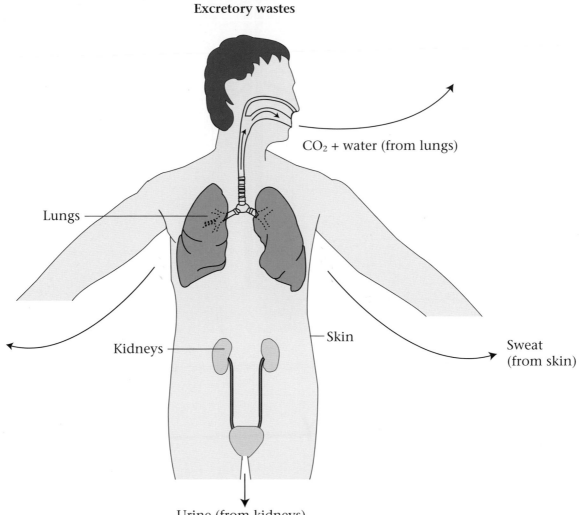

CO$_2$ + water (from lungs)

Lungs

Kidneys

Skin

Sweat
(from skin)

Urine (from kidneys)

Waste is produced by the skin, kidneys, and lungs		
Excretory organ	**What it excretes**	**Purpose**
Skin	Sweat	Cools the body.
Kidneys	Urine	Controls the water content of the body. Removes harmful urea.
Lungs	Carbon dioxide + water	These waste substances are produced by respiring cells.

Questions:
1. Which three organs produce excretory waste?
2. Where is sweat made?
3. Where is the excess water in our body removed?
4. What excretory wastes are produced in respiration?
5. When do we sweat a lot and why?

THE KIDNEYS

Functions of the kidneys
- Control the water level in the body.
- Remove harmful urea.

① **Aorta**
Carries oxygenated blood and waste to the kidneys.

② **Renal artery**
Carries oxygenated blood with waste into kidneys.

③ **Kidneys**
Urine is made in kidneys from waste materials in the blood. Kidneys use oxygen and glucose in respiration.

④ **Ureter**
This tube carries urine down to the bladder.

⑤ **Bladder**
Urine is stored here before removal.

⑥ **Urethra**
Urine passes through this tube and leaves the body.

⑧ **Vena cava**
Carries deoxygenated blood back to the heart.

A nephron

⑦ **Renal vein**
Carries blood with less waste away from the kidneys. The blood is deoxygenated.

Urine
Urine contains urea, excess water and salt.

SECTION THROUGH A KIDNEY

Nephrons
These are tiny tubes found in the kidneys.

Cortex

There are over a million nephrons in each kidney.

Medulla

Urine collects here from each nephron.

Renal artery

Renal vein

Urine

Renal artery
Blood entering the kidneys contains:
- Lots of oxygen.
- Lots of glucose
- Lots of water
- Lots of salt
- Lots of urea

} from food.

Renal vein
Blood leaving the kidneys contains:
- Less oxygen (some is respired).
- Less glucose (some is respired).
- Less urea (removed in urine).
- Right amount of water and salt (excess is removed in the urine).

Urea is a nitrogenous waste made from the breakdown of excess **protein**. It is **poisonous** if it builds up. Urea is made in the liver.

Questions:
1. Which blood vessel enters the kidneys?
2. Where is urine stored for a while?
3. Where is urine made?
4. What are the two functions of the kidneys?

A NEPHRON

What happens in the nephron?
- Blood is filtered.
- Useful materials are reabsorbed back into the blood.
- Waste is removed from the body as urine.

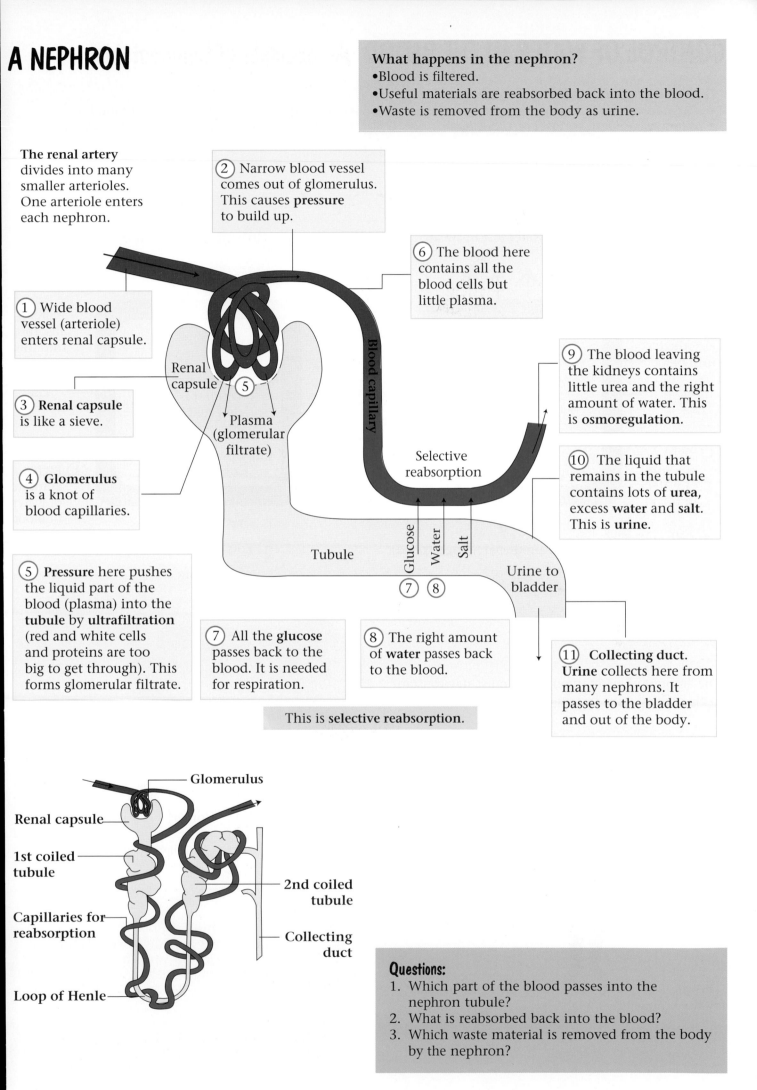

The renal artery divides into many smaller arterioles. One arteriole enters each nephron.

② Narrow blood vessel comes out of glomerulus. This causes **pressure** to build up.

⑥ The blood here contains all the blood cells but little plasma.

① Wide blood vessel (arteriole) enters renal capsule.

③ **Renal capsule** is like a sieve.

④ **Glomerulus** is a knot of blood capillaries.

⑨ The blood leaving the kidneys contains little urea and the right amount of water. This is **osmoregulation.**

⑩ The liquid that remains in the tubule contains lots of **urea,** excess **water** and **salt.** This is **urine.**

Renal capsule

⑤

Plasma (glomerular filtrate)

Blood capillary

Selective reabsorption

Tubule

Glucose

Water

Salt

⑦ ⑧

Urine to bladder

⑤ **Pressure** here pushes the liquid part of the blood (plasma) into the **tubule** by **ultrafiltration** (red and white cells and proteins are too big to get through). This forms glomerular filtrate.

⑦ All the **glucose** passes back to the blood. It is needed for respiration.

⑧ The right amount of **water** passes back to the blood.

⑪ **Collecting duct.** **Urine** collects here from many nephrons. It passes to the bladder and out of the body.

This is **selective reabsorption.**

Glomerulus

Renal capsule

1st coiled tubule

Capillaries for reabsorption

Loop of Henle

2nd coiled tubule

Collecting duct

Questions:
1. Which part of the blood passes into the nephron tubule?
2. What is reabsorbed back into the blood?
3. Which waste material is removed from the body by the nephron?

CONTROL OF WATER IN THE BLOOD An example of homeostasis.

How sweating affects the urine produced.
- The skin helps to control body temperature.
- When we are hot, we sweat to cool down.
- This reduces the water content of blood.
- Less water is lost in urine to conserve water.

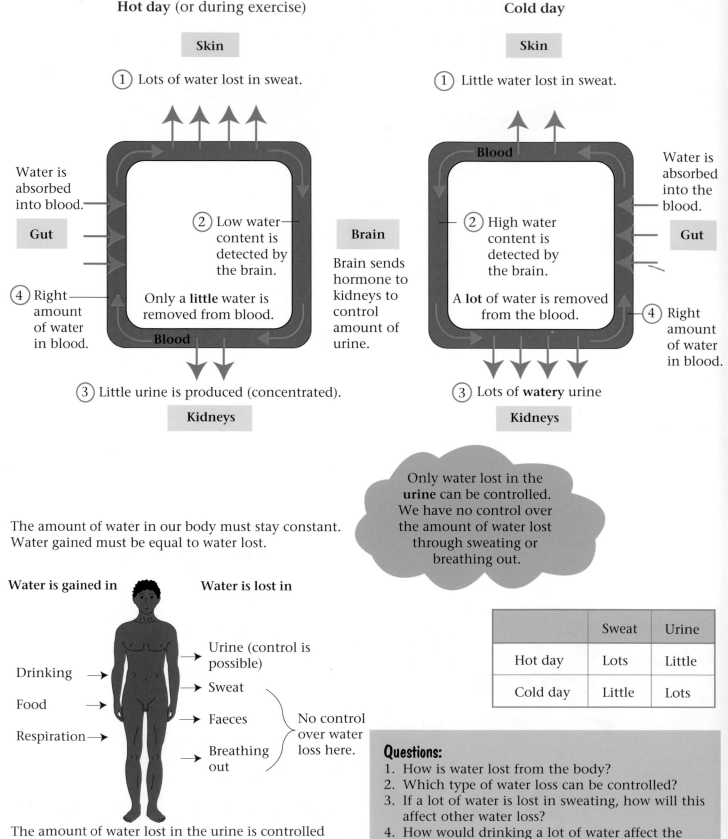

Hot day (or during exercise)

Skin

① Lots of water lost in sweat.

Water is absorbed into blood.

Gut

② Low water content is detected by the brain.

Only a **little** water is removed from blood.

Blood

④ Right amount of water in blood.

③ Little urine is produced (concentrated).

Kidneys

Cold day

Skin

① Little water lost in sweat.

Blood

② High water content is detected by the brain.

A **lot** of water is removed from the blood.

Water is absorbed into the blood.

Gut

④ Right amount of water in blood.

③ Lots of **watery** urine

Kidneys

Brain

Brain sends hormone to kidneys to control amount of urine.

Only water lost in the **urine** can be controlled. We have no control over the amount of water lost through sweating or breathing out.

The amount of water in our body must stay constant. Water gained must be equal to water lost.

Water is gained in   **Water is lost in**

Drinking →

Food →

Respiration →

→ Urine (control is possible)

→ Sweat

→ Faeces

→ Breathing out

No control over water loss here.

The amount of water lost in the urine is controlled by the **brain**.

	Sweat	Urine
Hot day	Lots	Little
Cold day	Little	Lots

Questions:
1. How is water lost from the body?
2. Which type of water loss can be controlled?
3. If a lot of water is lost in sweating, how will this affect other water loss?
4. How would drinking a lot of water affect the quantity of urine?
5. What is homeostasis?
6. What controls the level of water in the blood?

HORMONAL CONTROL OF WATER LEVEL

The **brain** releases **ADH** which controls the amount of **urine** produced (ADH = anti-diuretic hormone).

Less ADH means less water reabsorbed back into blood.

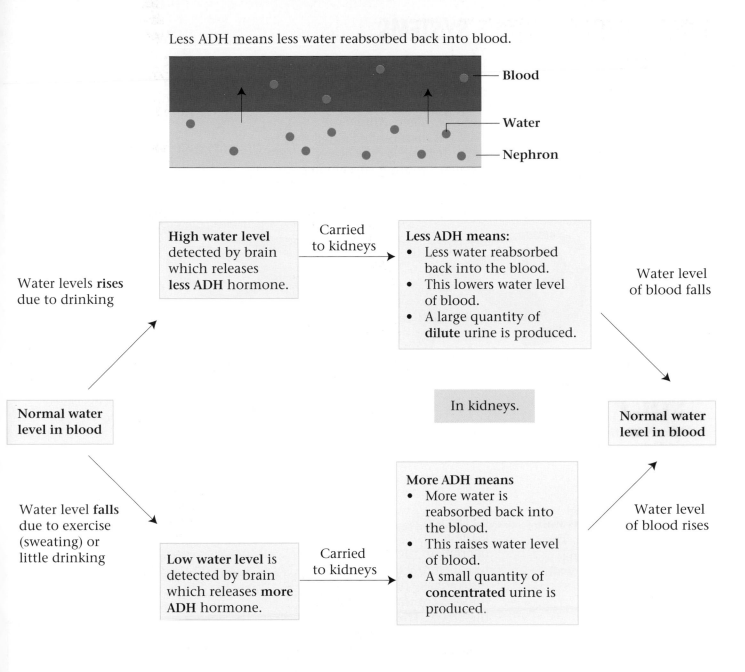

Water levels **rises** due to drinking

Normal water level in blood

High water level detected by brain which releases **less ADH** hormone.

Carried to kidneys

Less ADH means:
- Less water reabsorbed back into the blood.
- This lowers water level of blood.
- A large quantity of **dilute** urine is produced.

Water level of blood falls

In kidneys.

Normal water level in blood

Water level falls due to exercise (sweating) or little drinking

Low water level is detected by brain which releases **more ADH** hormone.

Carried to kidneys

More ADH means
- More water is reabsorbed back into the blood.
- This raises water level of blood.
- A small quantity of **concentrated** urine is produced.

Water level of blood rises

More ADH means more water reabsorbed back into blood.

Questions:
1. Where is the water level detected in the body?
2. Which hormone is produced to control the water level?
3. Where is the water level altered by this hormone?
4. On hot days do we produce more or less urine and why?

107

REPRODUCTION

HUMAN REPRODUCTIVE SYSTEMS

1. Female reproductive system

① Ovary
Here the hormones **oestrogen** and **progesterone** are produced. These cause the menstrual cycle and the secondary sexual characteristics. One ovum is released very 28 days from alternate ovaries. The release of the ovum is called **ovulation**.

⑥ The fertilised egg
is then pushed along the oviduct to the uterus to continue developing.

② The oviduct
The ovum from the ovary is pushed along the oviduct to the uterus. Cilia hairs sweep it along and muscles push the ovum along by peristalsis. The ovum lives for about one day.

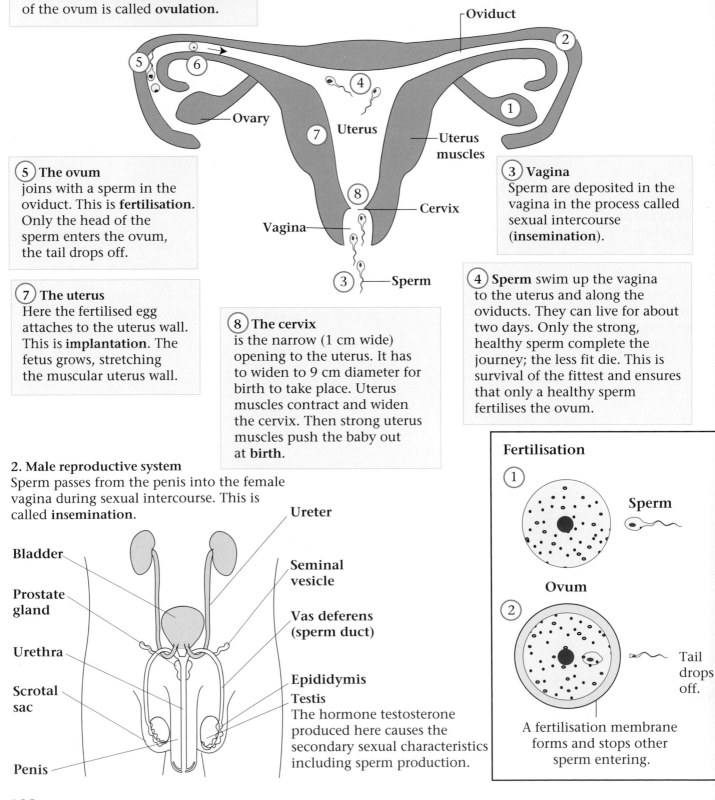

Oviduct

Ovary

Uterus

Uterus muscles

Cervix

Vagina

Sperm

⑤ The ovum
joins with a sperm in the oviduct. This is **fertilisation**. Only the head of the sperm enters the ovum, the tail drops off.

③ Vagina
Sperm are deposited in the vagina in the process called sexual intercourse (**insemination**).

④ Sperm swim up the vagina to the uterus and along the oviducts. They can live for about two days. Only the strong, healthy sperm complete the journey; the less fit die. This is survival of the fittest and ensures that only a healthy sperm fertilises the ovum.

⑦ The uterus
Here the fertilised egg attaches to the uterus wall. This is **implantation**. The fetus grows, stretching the muscular uterus wall.

⑧ The cervix
is the narrow (1 cm wide) opening to the uterus. It has to widen to 9 cm diameter for birth to take place. Uterus muscles contract and widen the cervix. Then strong uterus muscles push the baby out at **birth**.

2. Male reproductive system
Sperm passes from the penis into the female vagina during sexual intercourse. This is called **insemination**.

Bladder

Prostate gland

Urethra

Scrotal sac

Penis

Ureter

Seminal vesicle

Vas deferens (sperm duct)

Epididymis

Testis
The hormone testosterone produced here causes the secondary sexual characteristics including sperm production.

Fertilisation

① Sperm

Ovum

② Tail drops off.

A fertilisation membrane forms and stops other sperm entering.

108

THE MENSTRUAL CYCLE (28-day cycle).

This 28-day cycle is controlled by hormones. **One egg** is produced each cycle from alternate ovaries. The egg lives for only about one day and fertilisation is possible then. The uterus lining (**endometrium**) thickens up in preparation for a possible pregnancy. If fertilisation does not occur, this is shed from the body in the monthly period.

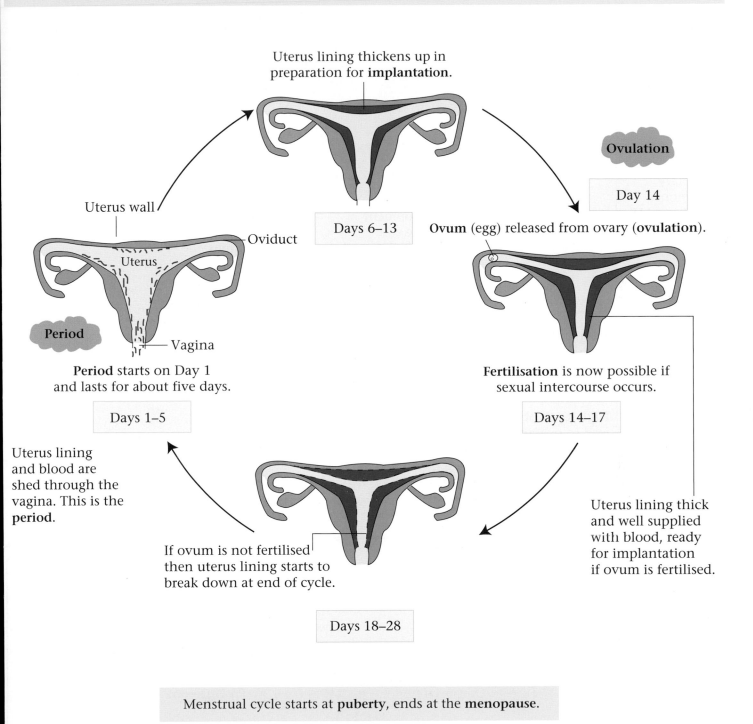

Uterus lining thickens up in preparation for **implantation**.

Ovulation

Day 14

Uterus wall

Oviduct

Uterus

Days 6–13

Ovum (egg) released from ovary (**ovulation**).

Period

Vagina

Period starts on Day 1 and lasts for about five days.

Days 1–5

Uterus lining and blood are shed through the vagina. This is the **period**.

Fertilisation is now possible if sexual intercourse occurs.

Days 14–17

Uterus lining thick and well supplied with blood, ready for implantation if ovum is fertilised.

If ovum is not fertilised then uterus lining starts to break down at end of cycle.

Days 18–28

Menstrual cycle starts at **puberty**, ends at the **menopause**.

Questions:
1. Where is the egg produced?
2. Where might fertilisation take place?
3. What is the purpose of the uterus lining thickening up?
4. What is ovulation?

HORMONAL CONTROL OF THE MENSTRUAL CYCLE

Changes in uterus lining (endometrium)	Breakdown of uterus lining. **Oestrogen** causes repair of uterus lining. **Progesterone** maintains the uterus lining. Thick uterus lining in readiness for implantation.
Changes in ovarian follicle	Ovum develops in the follicle. Ovum is released. **Corpus luteum** forms. Ovum. Follicle secretes oestrogen. Ovulation. C.L. secretes progesterone
Pituitary hormones (FSH and LH)	Peaks 1st. FSH. FSH causes development of the follicle containing the ovum. Peaks 3rd. LH. LH causes ovulation and the formation of the corpus luteum.
Ovarian hormones (oestrogen and progesterone)	**Oestrogen** and **progesterone** are produced by the ovaries from **puberty** to the **menopause** (from 12–50 years approximately). These cause the **secondary sexual characteristics** which change a girl into a woman, including the **menstrual cycle**. **High levels** of **oestrogen** lead to **ovulation**. Oestrogen is therefore found in **fertility drugs** to help a woman become pregnant. If an ovum is present, fertilisation and pregnancy may follow. Peaks 2nd. Oestrogen. Oestrogen repairs uterus lining and high levels stimulate the release of LH, causing ovulation. Peaks 4th. If pregnant. **Progesterone** maintains uterus lining and prepares the body for pregnancy.

1 Day 1 First day of period

7

14 (approx) Ovulation

21

28

Time in days

The human menstrual cycle is a 28 day cycle controlled by hormones. The hormones peak in the following order:

FSH ⟶ Oestrogen ⟶ LH ⟶ Progesterone

(FOLP)

The hormone **progesterone rises** when a woman is **pregnant**. It **stops ovulation** and keeps the uterus lining thick. As it stops ovulation, it is found in the **contraceptive pill** to stop unwanted pregnancies.

Questions:
1. How long is the menstrual cycle?
2. Which hormone repairs the uterus lining?
3. What is ovulation?
4. High levels of which hormone lead to ovulation?
5. What happens when both ovarian hormone levels fall?

THE PLACENTA Exchange of material between the mother and fetus.

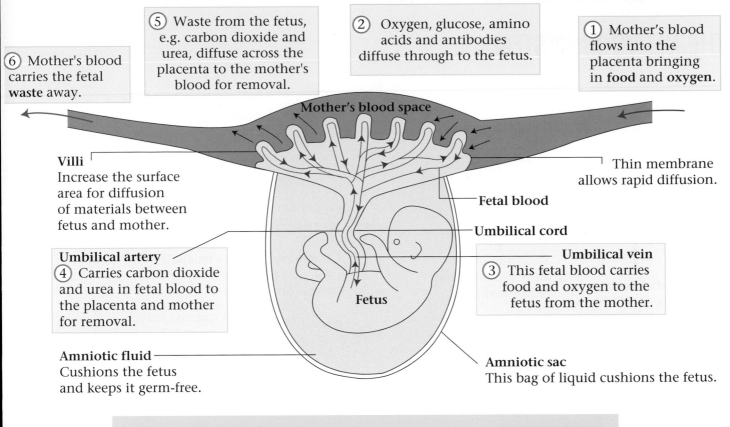

⑤ Waste from the fetus, e.g. carbon dioxide and urea, diffuse across the placenta to the mother's blood for removal.

② Oxygen, glucose, amino acids and antibodies diffuse through to the fetus.

① Mother's blood flows into the placenta bringing in **food** and **oxygen**.

⑥ Mother's blood carries the fetal **waste** away.

Mother's blood space

Villi
Increase the surface area for diffusion of materials between fetus and mother.

Thin membrane allows rapid diffusion.

Fetal blood

Umbilical cord

Umbilical artery
④ Carries carbon dioxide and urea in fetal blood to the placenta and mother for removal.

Umbilical vein
③ This fetal blood carries food and oxygen to the fetus from the mother.

Fetus

Amniotic fluid
Cushions the fetus and keeps it germ-free.

Amniotic sac
This bag of liquid cushions the fetus.

> **If the mother smokes**, this leads to low birthweight babies and premature births. Smoking deprives the baby of oxygen.

The **placenta** keeps the mother's and fetal blood *apart* while allowing exchange of materials. This separation is essential to:
- Stop the high pressure of the mother's blood destroying the fetal blood vessels.
- Stop the mother's blood rejecting the fetus .

Other substances that can cross the placenta to the fetus	
Substance	**Effect on baby**
Antibodies	Gives immunity to new born baby.
HIV	Causes AIDS.
Rubella virus (German measles)	Causes blindness, deafness, and brain damage.
*Nicotine	Increases heart rate.
*Carbon monoxide	Reduces oxygen reaching fetus.
(*From smoking)	

Thalidomide is a drug that was developed as a sleeping pill, but was also found to be effective at reducing morning sickness in pregnant women. Unfortunately, it was not tested before use, and many babies were born with severe limb abnormalities to mothers who took the drug in pregnancy. Many babies had no arms or legs. Recently, thalidomide is being used successfully to treat leprosy.

Questions:
1. Why must the mother's and fetal blood be kept apart?
2. Name two substances that pass from the mother to the fetus.
3. How does the fetus get rid of waste?
4. What features of the placenta increase the rate of diffusion?

COORDINATION

THE NERVOUS SYSTEM

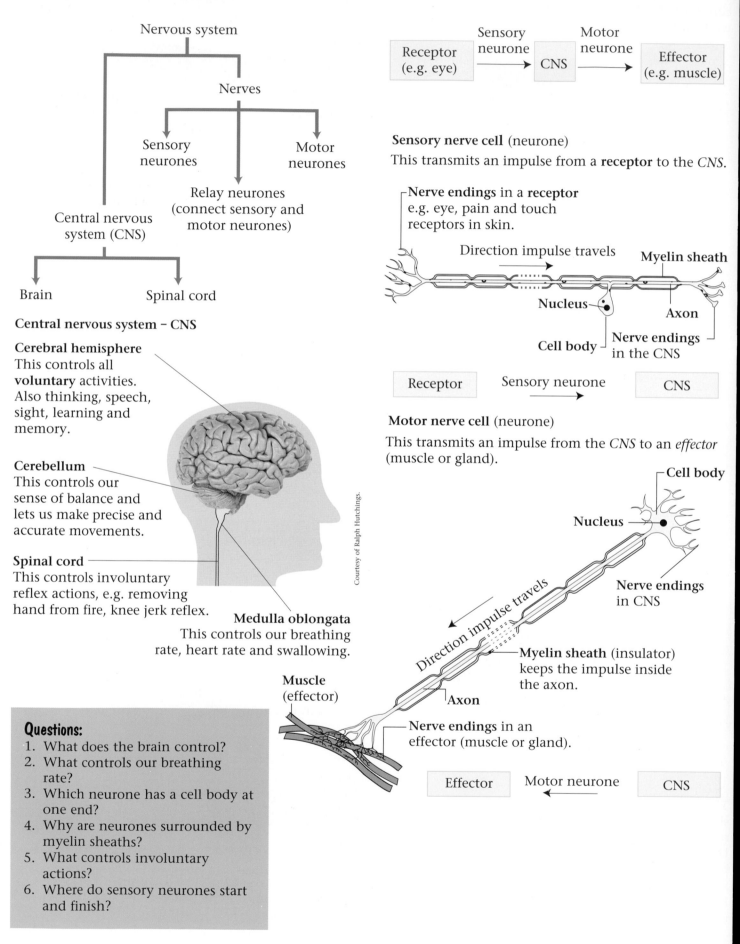

Nervous system
→ Nerves
  → Sensory neurones
  → Relay neurones (connect sensory and motor neurones)
  → Motor neurones
→ Central nervous system (CNS)
  → Brain
  → Spinal cord

Central nervous system – CNS

Cerebral hemisphere
This controls all **voluntary** activities. Also thinking, speech, sight, learning and memory.

Cerebellum
This controls our sense of balance and lets us make precise and accurate movements.

Spinal cord
This controls involuntary reflex actions, e.g. removing hand from fire, knee jerk reflex.

Medulla oblongata
This controls our breathing rate, heart rate and swallowing.

Courtesy of Ralph Hutchings.

Receptor (e.g. eye) → Sensory neurone → CNS → Motor neurone → Effector (e.g. muscle)

Sensory nerve cell (neurone)
This transmits an impulse from a **receptor** to the *CNS*.

Nerve endings in a **receptor** e.g. eye, pain and touch receptors in skin.

Direction impulse travels

Myelin sheath

Nucleus

Axon

Cell body

Nerve endings in the CNS

Receptor → Sensory neurone → CNS

Motor nerve cell (neurone)
This transmits an impulse from the *CNS* to an *effector* (muscle or gland).

Cell body

Nucleus

Nerve endings in CNS

Direction impulse travels

Myelin sheath (insulator) keeps the impulse inside the axon.

Axon

Muscle (effector)

Nerve endings in an effector (muscle or gland).

Effector ← Motor neurone ← CNS

Questions:
1. What does the brain control?
2. What controls our breathing rate?
3. Which neurone has a cell body at one end?
4. Why are neurones surrounded by myelin sheaths?
5. What controls involuntary actions?
6. Where do sensory neurones start and finish?

THE SPINAL CORD This controls reflex actions, e.g. knee jerk, withdrawal of hand from fire.

1) Hand in fire stimulates pain **receptor** in skin.

2) Message is sent as an impulse along the **sensory neurone** to the spinal cord.

3) Sensory neurone is found in the **dorsal root** of spinal nerve.

4) Message passes by diffusion of chemicals across the **synapse**. This causes an impulse to pass to the **relay neurone**.

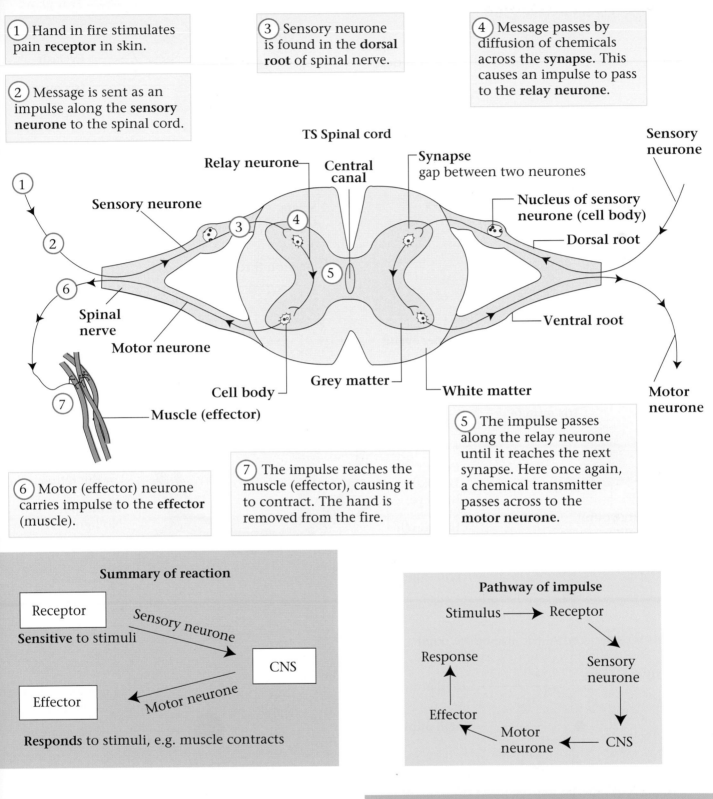

TS Spinal cord

Relay neurone — Central canal

Synapse gap between two neurones

Sensory neurone

Sensory neurone

Nucleus of sensory neurone (cell body)

Dorsal root

Spinal nerve

Motor neurone

Ventral root

Cell body — Grey matter — White matter

Motor neurone

Muscle (effector)

5) The impulse passes along the relay neurone until it reaches the next synapse. Here once again, a chemical transmitter passes across to the **motor neurone**.

6) Motor (effector) neurone carries impulse to the **effector** (muscle).

7) The impulse reaches the muscle (effector), causing it to contract. The hand is removed from the fire.

Summary of reaction

Receptor — **Sensitive to stimuli**

Sensory neurone → CNS

Effector ← Motor neurone

Responds to stimuli, e.g. muscle contracts

Pathway of impulse

Stimulus → Receptor

Sensory neurone

Response

Effector ← Motor neurone ← CNS

Questions:
1. What does the spinal cord control?
2. What is stimulated by pain?
3. How is a message sent along to the spinal cord?
4. What is a synapse and how do messages cross this?
5. Which neurone carries an impulse to the effector?
6. What is an effector?

Definition:
A **reflex action** is a rapid, automatic response to a stimulus. These involuntary actions are usually controlled by the spinal cord.

113

THE EYE

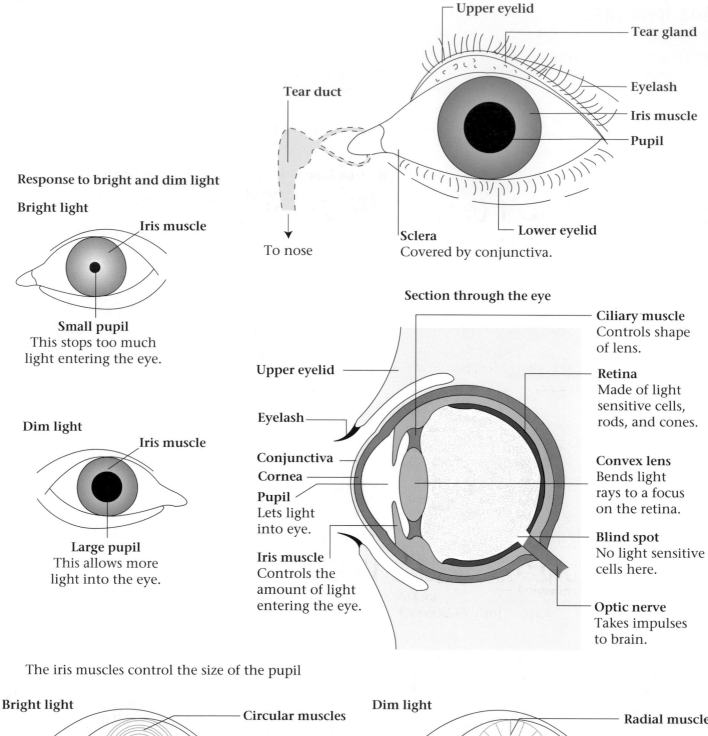

Upper eyelid

Tear gland

Eyelash

Iris muscle

Pupil

Tear duct

To nose

Sclera
Covered by conjunctiva.

Lower eyelid

Response to bright and dim light

Bright light

Iris muscle

Small pupil
This stops too much
light entering the eye.

Dim light

Iris muscle

Large pupil
This allows more
light into the eye.

Section through the eye

Ciliary muscle
Controls shape
of lens.

Upper eyelid

Eyelash

Conjunctiva
Cornea

Pupil
Lets light
into eye.

Iris muscle
Controls the
amount of light
entering the eye.

Retina
Made of light
sensitive cells,
rods, and cones.

Convex lens
Bends light
rays to a focus
on the retina.

Blind spot
No light sensitive
cells here.

Optic nerve
Takes impulses
to brain.

The iris muscles control the size of the pupil

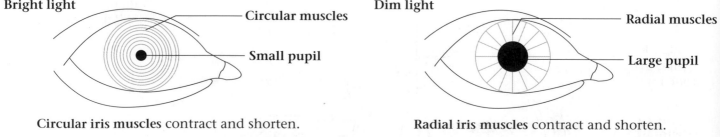

Bright light

Circular muscles

Small pupil

Circular iris muscles contract and shorten.

Dim light

Radial muscles

Large pupil

Radial iris muscles contract and shorten.

Questions:

1. Which part of the eye is made of light sensitive cells?
2. What type of lens is found in the eye and what does it do to light rays?
3. Which nerve takes impulses from the eye to the brain?
4. Which muscles control the size of the pupil?
5. What size is the pupil in bright conditions and why?
6. Which muscle can control the shape of the lens in the eye?

VISION - HOW WE SEE

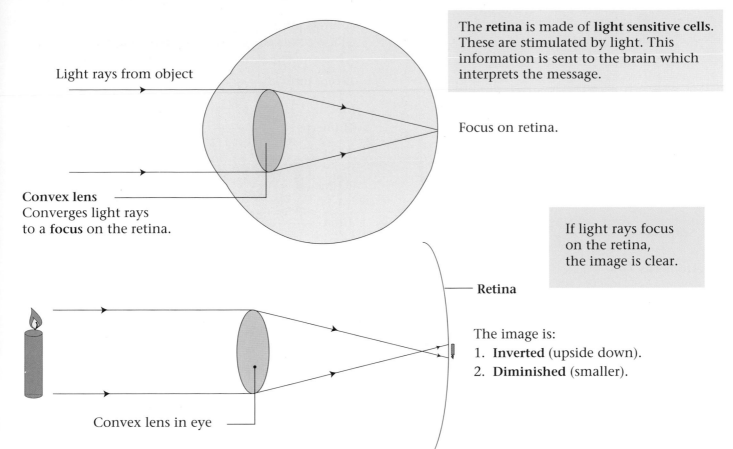

Light rays from object

Convex lens
Converges light rays
to a **focus** on the retina.

The **retina** is made of **light sensitive cells.** These are stimulated by light. This information is sent to the brain which interprets the message.

Focus on retina.

If light rays focus on the retina, the image is clear.

— Retina

The image is:
1. **Inverted** (upside down).
2. **Diminished** (smaller).

Convex lens in eye

Convex lens refracts or bends light to a focus on the retina.

Object	Light rays entering eye	Ciliary muscle	Lens shape	Effect
Near	Light rays are diverging	Contracts	Fatter F	**Converges** the **diverging** light rays onto the **retina** to give a **clear image.**
Distant	Light rays are parallel	Relaxes	Thinner F	**Converges** the **parallel** light rays onto the **retina** to give a **clear image.**

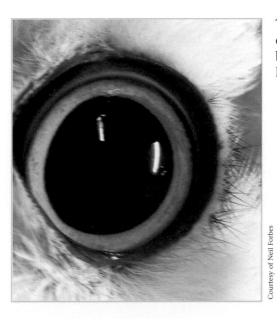

Courtesy of Neil Forbes

The large eyes of an owl provide excellent vision for finding and catching prey such as mice. Their forward facing eyes provide binocular vision enabling them to judge distance accurately for hunting.

Questions:
1. What shape lens is needed for looking at near objects?
2. What effect does the lens have on light rays entering the eye?
3. Which muscle controls the shape of the lens?

SKELETON AND MOVEMENT

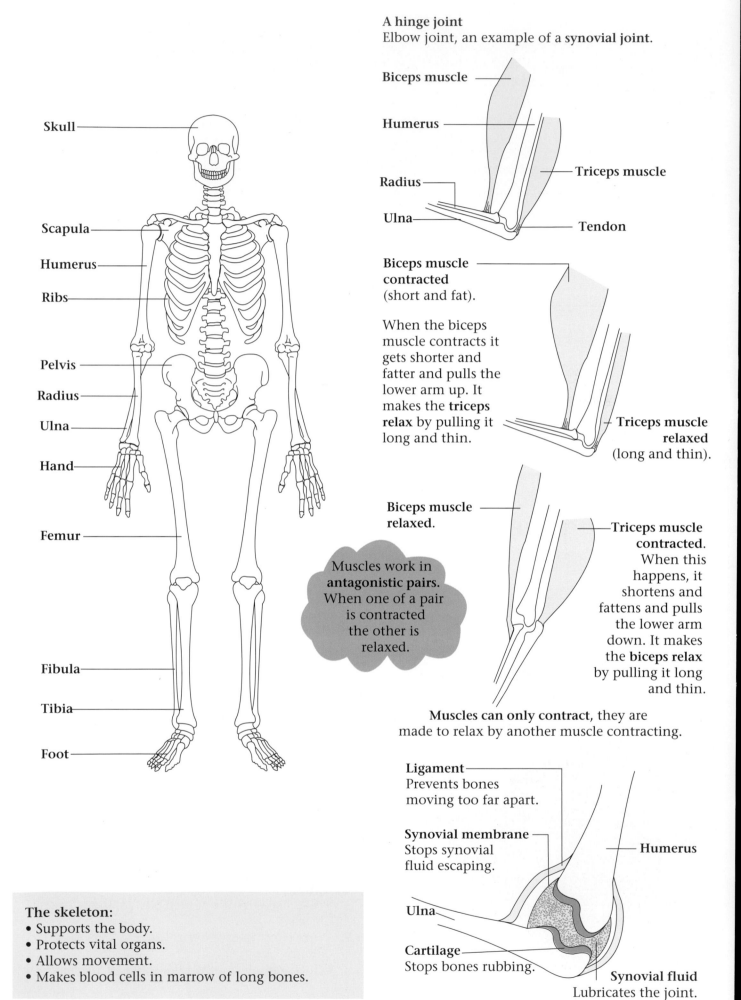

Skull

Scapula

Humerus

Ribs

Pelvis

Radius

Ulna

Hand

Femur

Fibula

Tibia

Foot

A hinge joint
Elbow joint, an example of a **synovial joint**.

Biceps muscle

Humerus

Radius

Ulna

Triceps muscle

Tendon

Biceps muscle
contracted
(short and fat).

When the biceps
muscle contracts it
gets shorter and
fatter and pulls the
lower arm up. It
makes the **triceps
relax** by pulling it
long and thin.

Triceps muscle
relaxed
(long and thin).

Biceps muscle
relaxed.

Muscles work in
antagonistic pairs.
When one of a pair
is contracted
the other is
relaxed.

Triceps muscle
contracted.
When this
happens, it
shortens and
fattens and pulls
the lower arm
down. It makes
the **biceps relax**
by pulling it long
and thin.

Muscles can only contract, they are
made to relax by another muscle contracting.

Ligament
Prevents bones
moving too far apart.

Synovial membrane
Stops synovial
fluid escaping.

Humerus

Ulna

Cartilage
Stops bones rubbing.

Synovial fluid
Lubricates the joint.

The skeleton:
• Supports the body.
• Protects vital organs.
• Allows movement.
• Makes blood cells in marrow of long bones.

PLANTS

LEAVES AND PHOTOSYNTHESIS Photosynthesis is the process by which plants make sugar.

Equation of photosynthesis

$$\text{Water + carbon dioxide} \xrightarrow[\text{Light}]{\text{Chlorophyll}} \text{Sugar + oxygen}$$

$$6H_2O + 6CO_2 \longrightarrow C_6H_{12}O_6 + 6O_2$$

Photosynthesis takes place only during the day as light is needed.

Light

Water + CO_2 Light Chlorophyll Sugar + O_2

O_2

Water from roots

CO_2

CO_2

Stomata

Carbon dioxide from air enters through stomata.

How leaves are adapted for photosynthesis

- Leaves are thin and flat to trap more light.
- Upper layers are transparent to let light through to palisade layer.
- Palisade layer has lots of chloroplasts to trap light.
- Palisade cells are arranged so that the maximum number receive light.
- Leaves can turn to face the light.
- Veins bring in water for photosynthesis.
- Stomata allow entry of carbon dioxide.
- Spongy mesophyll allows gases to circulate.

Leaf structure TS

Waxy cuticle
Waterproof layer stops leaf drying up.

Palisade layer
Here many chloroplasts absorb light for photosynthesis. Maximum number of cells can get light.

Vein Xylem phloem

Air space

Upper epidermis (protection)

Chloroplasts
Absorb light for photosynthesis.

Spongy mesophyll
Allows gases to circulate.

Lower epidermis has many stomata.

Stomata
Let gases in and out.

Guard cells Control opening and closing of stomata.

Questions:
1. What is photosynthesis?
2. In which part of the plant does photosynthesis occur?
3. Why does photosynthesis take place only during the day?
4. Carbon dioxide is needed for photosynthesis. How does it enter the plant?
5. In which layer of the leaf does most photosynthesis take place and why?
6. Describe three ways in which leaves are suited for photosynthesis.

TRANSPORT IN PLANTS

Photosynthesis takes place mainly in the **chloroplasts** of the **palisade layer** in the leaf.

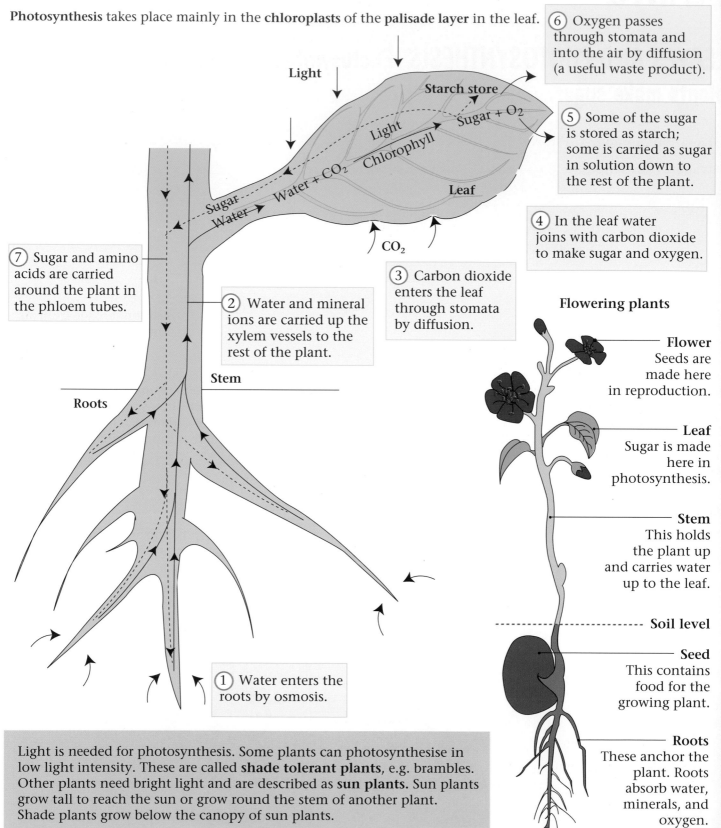

⑥ Oxygen passes through stomata and into the air by diffusion (a useful waste product).

Light

Starch store

Light

Chlorophyll

Sugar + O₂

Sugar

Water + CO₂

Water

Leaf

CO₂

⑤ Some of the sugar is stored as starch; some is carried as sugar in solution down to the rest of the plant.

④ In the leaf water joins with carbon dioxide to make sugar and oxygen.

⑦ Sugar and amino acids are carried around the plant in the phloem tubes.

② Water and mineral ions are carried up the xylem vessels to the rest of the plant.

③ Carbon dioxide enters the leaf through stomata by diffusion.

Stem

Roots

① Water enters the roots by osmosis.

Flowering plants

Flower
Seeds are made here in reproduction.

Leaf
Sugar is made here in photosynthesis.

Stem
This holds the plant up and carries water up to the leaf.

Soil level

Seed
This contains food for the growing plant.

Roots
These anchor the plant. Roots absorb water, minerals, and oxygen.

Light is needed for photosynthesis. Some plants can photosynthesise in low light intensity. These are called **shade tolerant plants**, e.g. brambles. Other plants need bright light and are described as **sun plants.** Sun plants grow tall to reach the sun or grow round the stem of another plant. Shade plants grow below the canopy of sun plants.

Questions:
1. Where is sugar made?
2. How is this sugar carried to other parts of the plant?
3. Water is needed for photosynthesis. How is it carried to the leaves?
4. Which gas is produced as a waste product in photosynthesis and what happens to it?
5. What happens to the sugar that stays in the leaf?
6. What two materials are taken into the leaf for photosynthesis?
7. In the stem, xylem and phloem are found. What do they each carry and in what direction?

USES OF SUGAR MADE IN PHOTOSYNTHESIS

Sugar made in the leaves is carried all over the plant in the phloem tubes. This is possible as sugar is **soluble**. This transport of sugar is called **translocation**.

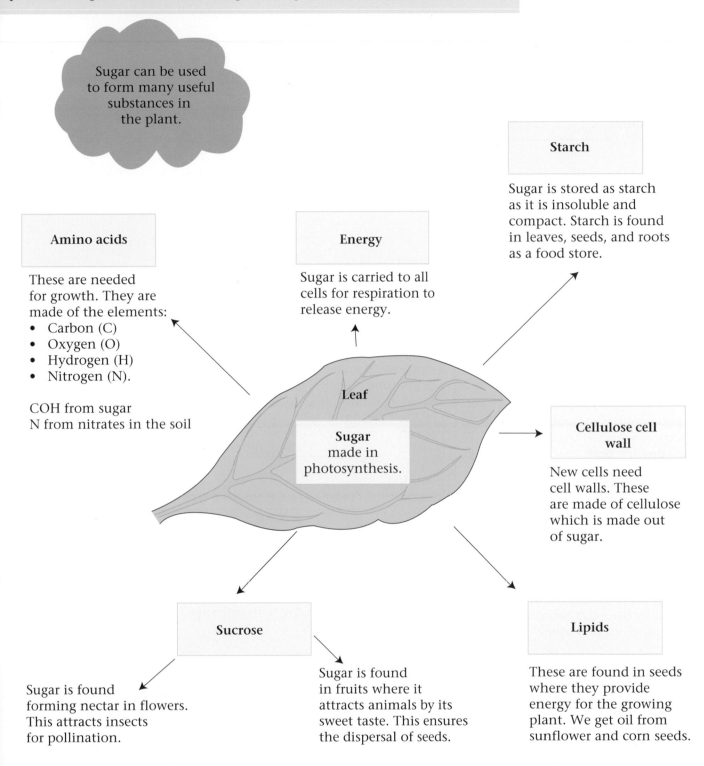

Sugar can be used to form many useful substances in the plant.

Starch

Sugar is stored as starch as it is insoluble and compact. Starch is found in leaves, seeds, and roots as a food store.

Amino acids

These are needed for growth. They are made of the elements:
- Carbon (C)
- Oxygen (O)
- Hydrogen (H)
- Nitrogen (N).

COH from sugar
N from nitrates in the soil

Energy

Sugar is carried to all cells for respiration to release energy.

Leaf

Sugar made in photosynthesis.

Cellulose cell wall

New cells need cell walls. These are made of cellulose which is made out of sugar.

Sucrose

Sugar is found forming nectar in flowers. This attracts insects for pollination.

Sugar is found in fruits where it attracts animals by its sweet taste. This ensures the dispersal of seeds.

Lipids

These are found in seeds where they provide energy for the growing plant. We get oil from sunflower and corn seeds.

Questions:
1. What is translocation? In what tube does this occur?
2. Every cell in the plant needs sugar for respiration. What is released in this process?
3. Why is sucrose needed in the flower? Give two reasons.
4. How is sugar stored and why?
5. What two materials might sugar be changed to in the seeds and why?
6. Why is sugar easy to transport round the plant?
7. To grow, plants need amino acids. What element is added to sugar to provide this? How do plants take in this element?

LIMITING FACTORS IN PHOTOSYNTHESIS

Photosynthesis is a process that needs **light**, **warmth** and **carbon dioxide**. If one of these factors is in **short supply**, it will limit the rate of photosynthesis, and so is called a **limiting factor**.

1. Light as limiting factor

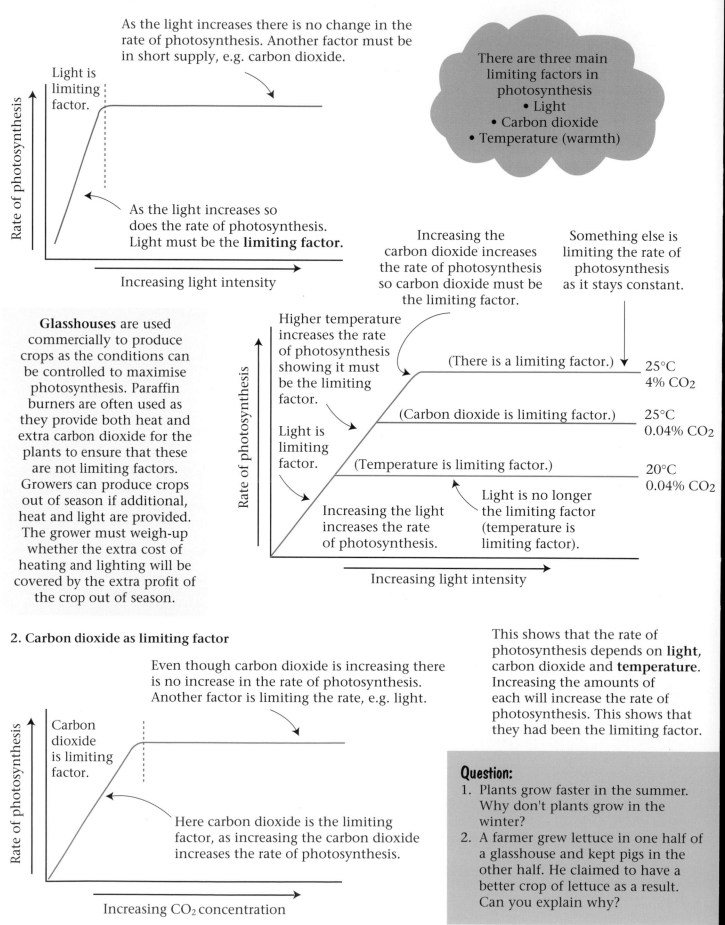

As the light increases there is no change in the rate of photosynthesis. Another factor must be in short supply, e.g. carbon dioxide.

Light is limiting factor.

As the light increases so does the rate of photosynthesis. Light must be the **limiting factor**.

Rate of photosynthesis

Increasing light intensity

There are three main limiting factors in photosynthesis
- Light
- Carbon dioxide
- Temperature (warmth)

Glasshouses are used commercially to produce crops as the conditions can be controlled to maximise photosynthesis. Paraffin burners are often used as they provide both heat and extra carbon dioxide for the plants to ensure that these are not limiting factors. Growers can produce crops out of season if additional, heat and light are provided. The grower must weigh-up whether the extra cost of heating and lighting will be covered by the extra profit of the crop out of season.

Increasing the carbon dioxide increases the rate of photosynthesis so carbon dioxide must be the limiting factor.

Something else is limiting the rate of photosynthesis as it stays constant.

Higher temperature increases the rate of photosynthesis showing it must be the limiting factor.

Light is limiting factor.

Increasing the light increases the rate of photosynthesis.

(There is a limiting factor.) — $25°C$ 4% CO_2

(Carbon dioxide is limiting factor.) — $25°C$ 0.04% CO_2

(Temperature is limiting factor.) — $20°C$ 0.04% CO_2

Light is no longer the limiting factor (temperature is limiting factor).

Rate of photosynthesis

Increasing light intensity

2. Carbon dioxide as limiting factor

Even though carbon dioxide is increasing there is no increase in the rate of photosynthesis. Another factor is limiting the rate, e.g. light.

Carbon dioxide is limiting factor.

Here carbon dioxide is the limiting factor, as increasing the carbon dioxide increases the rate of photosynthesis.

Rate of photosynthesis

Increasing CO_2 concentration

This shows that the rate of photosynthesis depends on **light**, carbon dioxide and **temperature**. Increasing the amounts of each will increase the rate of photosynthesis. This shows that they had been the limiting factor.

Question:
1. Plants grow faster in the summer. Why don't plants grow in the winter?
2. A farmer grew lettuce in one half of a glasshouse and kept pigs in the other half. He claimed to have a better crop of lettuce as a result. Can you explain why?

MINERALS AND PLANTS

Plants make sugar in photosynthesis. To stay healthy, they also need minerals.

These cuttings are all the same species, same age and started off the same size.

Experiment with cuttings of *Tradescantia* grown in solution.

Small leaves

Weak stem

Poor root growth

No nitrogen
This plant has no nitrogen and is unable to make protein for growth.

Plants normally take in nitrogen as nitrates which are soluble and enter roots dissolved in water.

Green leaves

Strong stem

Good root growth

All minerals
Plant grown in solution with all minerals required for healthy growth.

These are taken in through the roots dissolved in water.

Yellow, small leaves

Weak stem

Poor root growth

No magnesium
Magnesium is needed to make chlorophyll (green pigment).

Without chlorophyll, little sugar can be made, so little growth.

Uptake of minerals

Minerals can enter by **diffusion** and **active transport** (low to high concentration).

Nitrates in

High nitrate conc.

Low nitrate concentration in soil

Nitrates in

Root hair cell

Nitrates move from the lower concentration in the soil to the higher concentration in the root cells. This is the opposite of diffusion and requires energy. It is called active transport.

Element/Mineral	Why it is needed	Deficiency causes
Nitrates	For proteins and growth.	Poor growth.
Magnesium	To make chlorophyll.	Yellow leaves.
Iron	To make chlorophyll.	Yellow leaves.
Potassium	For photosynthesis and respiration.	Yellow-edged leaves.
Phosphate	To make protein and for cell membranes.	Stunted growth of roots.

Questions:
1. Where do minerals enter a plant?
2. By what two processes can minerals enter?
3. If a plant is grown without magnesium, how will it be affected?
4. Why do plants need nitrogen? How does it enter the plant, in what form?
5. Why is it important that the same plant was used in this experiment at the same size and age?
6. How does energy help in the uptake of minerals and how do plants release energy?

WATER MOVEMENT THROUGH A PLANT

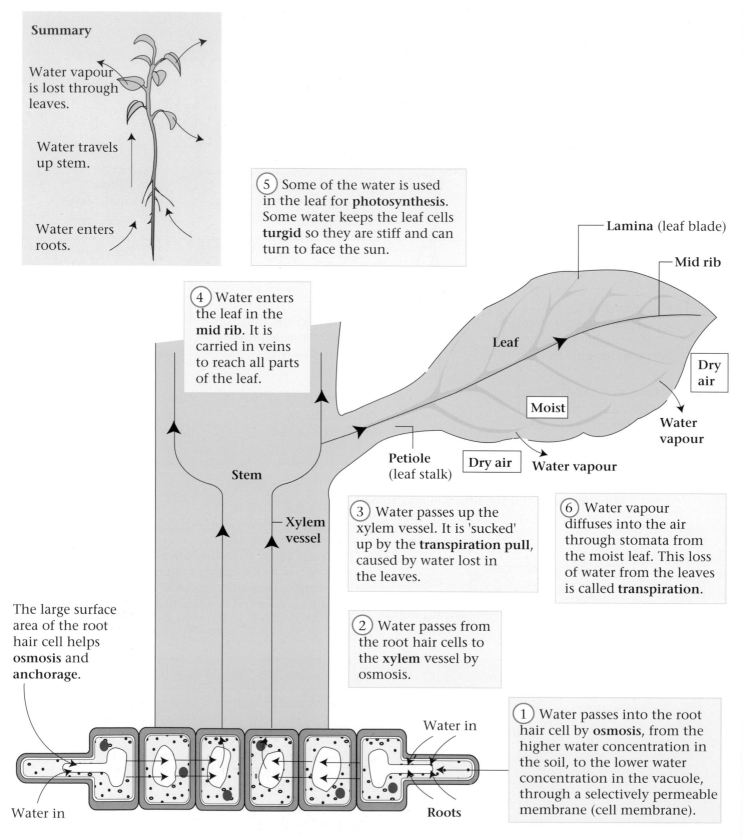

Summary

Water vapour is lost through leaves.

Water travels up stem.

Water enters roots.

5 Some of the water is used in the leaf for **photosynthesis**. Some water keeps the leaf cells **turgid** so they are stiff and can turn to face the sun.

Lamina (leaf blade)

Mid rib

Leaf

Dry air

Water vapour

Moist

4 Water enters the leaf in the **mid rib**. It is carried in veins to reach all parts of the leaf.

Petiole (leaf stalk)

Dry air Water vapour

Stem

Xylem vessel

3 Water passes up the xylem vessel. It is 'sucked' up by the **transpiration pull**, caused by water lost in the leaves.

6 Water vapour diffuses into the air through stomata from the moist leaf. This loss of water from the leaves is called **transpiration**.

2 Water passes from the root hair cells to the **xylem** vessel by osmosis.

The large surface area of the root hair cell helps **osmosis** and **anchorage**.

Water in

Water in

Roots

1 Water passes into the root hair cell by **osmosis**, from the higher water concentration in the soil, to the lower water concentration in the vacuole, through a selectively permeable membrane (cell membrane).

Questions:
1. How does the shape of the root hair cell help in the uptake of water?
2. Why does water enter the plant roots and how does rain help this process?
3. Which vessel carries water up the plant?
4. How does water enter the leaf?
5. How is water used in the leaf?
6. How does water vapour pass out of the leaf and into the air?
7. What is this loss of water called?
8. What is the transpiration pull?

TRANSPIRATION   The loss of water from a plant.

Diffusion

Few stomata on upper surface.

Wet

Water vapour

Many stomata on lower surface to reduce water loss.

Water vapour diffuses out.

Dry air

Importance of transpiration
Water is lost from the leaves in transpiration, through holes called stomata.
- Water *with minerals* enters roots to replace water lost through leaves.
- Nitrates and phosphates are vital for healthy plant growth.
- Water is needed for turgidity and support in plants; without water plants wilt.
- Water is needed for photosynthesis and translocation.
- Loss of water helps to cool plants down.

Why transpiration occurs
Stomata need to be open to let gases in and out for photosynthesis and respiration. Unfortunately, water vapour can escape from the leaf through these holes. On hot, dry and windy days many plants close some of their stomata to reduce this loss of water. The opening and closing of stomata is controlled by the **guard cells**.

Comparison of water loss in different conditions

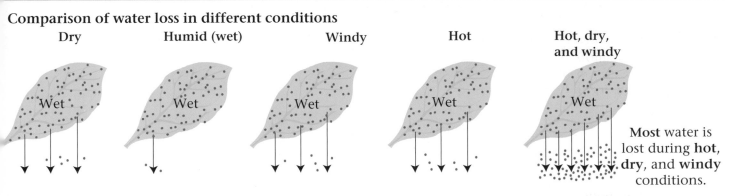

Dry

Humid (wet)

Windy

Hot

Hot, dry, and windy

Wet

Most water is lost during **hot**, **dry**, and **windy** conditions.

Holes in the leaf are called **stomata**. If it is moist in the leaf and dry outside, water vapour comes out by **diffusion**. **The greater the difference in concentration inside and outside the leaf, the faster the water loss by diffusion.** More water comes from the roots to replace water lost in the leaves.

Water vapour diffuses out and forms a **diffusion shell** of damp air. This layer slows the rate of diffusion as the difference in concentration of water vapour inside and outside the leaf is reduced. These 'diffusion shells' build up only on still days, as the wind blows any diffusion shells away, causing a rapid rate of diffusion.

Most stomata are found on the **lower surface** of the leaf. Here it is cooler and less windy than on the upper surface. This reduces water loss from the leaf.

At night stomata close. This reduces water loss in transpiration.

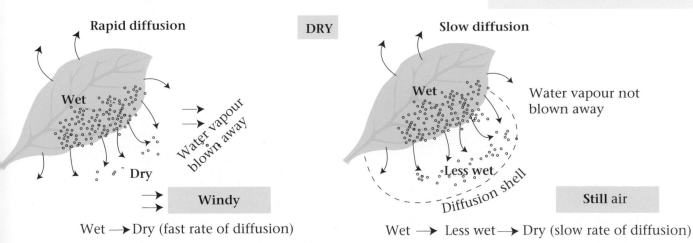

Rapid diffusion

DRY

Slow diffusion

Wet

Water vapour blown away

Dry

Windy

Wet → Dry (fast rate of diffusion)

Wet

Water vapour not blown away

Less wet

Diffusion shell

Still air

Wet → Less wet → Dry (slow rate of diffusion)

OPENING AND CLOSING OF STOMATA

1. Lower surface of leaf

Stoma
Lets gases in and out.

Lower epidermal cells

Guard cell

Vacuole (cell sap). Contains water and solutes, gives cell support.

Vacuole
Contains cell sap.

Thick inner cellulose cell wall

Chloroplast
Photosynthesis here.

No chloroplasts
Lower epidermal cells do not photosynthesise.

2. Plant cells

When water enters

Water in

Cellulose cell wall

Swollen vacuole

This is **turgid**.

When water leaves

Water out

The cell membrane starts to lose contact with the cell wall.

Small vacuole

This is **flaccid**.

Plant cells full of water are firm and turgid. This gives them support.

3. Guard cells

Day

Water enters by **osmosis**

Cell sap stronger

Vacuole

Chloroplasts

Stoma open

Vacuole swells

Turgid

Water enters guard cells from surrounding cells by osmosis and the vacuoles swells.

↓

Guard cells swell up and become turgid.

↓

This pulls the guard cells apart.

↓

Stomata open.

Night

Stoma closed

Water leaves by osmosis

Cell sap weaker

Vacuole shrinks

Flaccid

Water leaves guard cells by osmosis and the vacuole shrinks.

↓

Guard cells are floppy, flaccid.

↓

The guard cells collapse together.

↓

Stomata close.

Questions:
1. Which cells on the lower surface of the leaf are able to photosynthesise and why?
2. What do we call plant cells that contain a lot of water?
3. How does the entry of water to guard cells affect the stomata?
4. What stops plant cells bursting when water enters?

LEAVES

1. Marram grass – a **xerophyte** (a plant adapted to dry conditions). Found in **dry (arid)** conditions, i.e. sand dunes.

Leaf is **curled** to reduce area exposed to sun and wind, to reduce water loss.

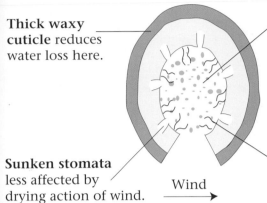

Thick waxy cuticle reduces water loss here.

Water vapour trapped.

Sunken stomata less affected by drying action of wind.

Wind

Hairs reduce air movement in leaf, reducing the drying action of wind.

Contrasting leaves

Marram	Beech
Thick cuticle	Thin cuticle
Leaf curled	Leaf flat
Sunken stomata	Stomata at surface
Little surface area exposed	Large surface area exposed
Hairs to reduce air movement	No hairs
Designed to reduce water loss in dry conditions	Designed for maximum photosynthesis

Other adaptations to living in dry conditions
- Extensive roots to reach any available water.
- Storage of water in succulent stem.
- Stomata which close in dry conditions.
- Thick waxy cuticle in stem to reduce water loss.

1. Beech leaf – found in **damp** conditions, i.e. woodland.

Veins
Carry water to all parts of leaf and take sugar away.

Mid rib
Supports the leaf.

Leaf blade (lamina)
Large and flat and faces the sun to absorb light for photosynthesis. Has lots of stomata to let gases in and out.

Leaf stalk
Contains veins bringing water in and taking sugar to rest of plant.

Leaves at night Only respiration as light is not available.

Respiration
Oxygen + sugar $\longrightarrow$ Carbon dioxide + water + energy

└─from air └─from sugar made during day in photosynthesis

CO_2

O_2 Respiration

Leaves in day Both respiration and photosynthesis occur.

Respiration
Oxygen + sugar $\longrightarrow$ Carbon dioxide + water + energy

Photosynthesis
Carbon dioxide + water $\xrightarrow[\text{Chlorophyll}]{\text{Light}}$ Sugar + oxygen

Overall exchange
O_2
CO_2

Respiration
CO_2
O_2

CO_2
O_2
Photosynthesis

Questions:
1. Name a plant that lives in dry conditions.
2. What features of the plant reduce water loss from the leaves?
3. What other features are found in plants living in dry conditions?
4. Beech leaves are found in damp areas. What structures in the leaf allow water to reach every part?
5. What gas enters a leaf at night? Which process is taking place in the leaf.
6. What processes take place in a leaf during the day?
7. Overall, during the day, what gases a) enter the leaf, b) pass out of the leaf?

LIFE CYCLE OF A PLUM TREE

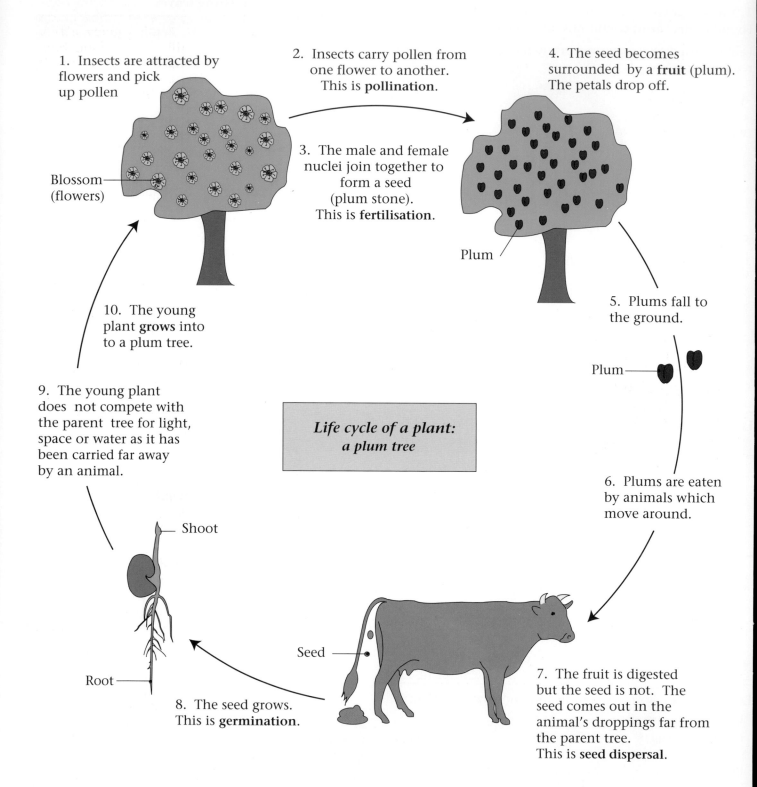

1. Insects are attracted by flowers and pick up pollen

2. Insects carry pollen from one flower to another. This is **pollination**.

3. The male and female nuclei join together to form a seed (plum stone). This is **fertilisation**.

4. The seed becomes surrounded by a **fruit** (plum). The petals drop off.

Blossom (flowers)

Plum

5. Plums fall to the ground.

Plum

Life cycle of a plant: a plum tree

6. Plums are eaten by animals which move around.

10. The young plant **grows** into to a plum tree.

9. The young plant does not compete with the parent tree for light, space or water as it has been carried far away by an animal.

Shoot

Root

7. The fruit is digested but the seed is not. The seed comes out in the animal's droppings far from the parent tree. This is **seed dispersal**.

Seed

8. The seed grows. This is **germination**.

Questions:
1. What is germination?
2. What is pollination?
3. What forms after fertilisation?
4. How do plums help in seed dispersal?
5. Why is it important that seeds are dispersed away from the parent plant?

FLOWERS AND REPRODUCTION

1. **Petals**. Colour and scent attract insects, which are rewarded with *nectar* (a sugar solution in the nectary). As insect searches for nectar, pollen is rubbed onto its body, which is carried to another flower. This is **pollination**.

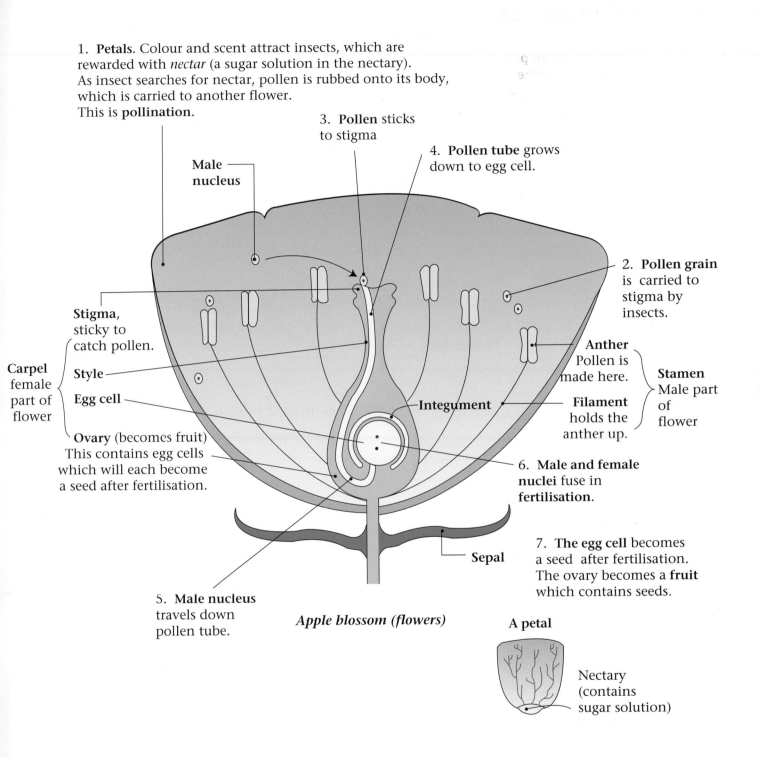

Male nucleus

3. **Pollen** sticks to stigma

4. **Pollen tube** grows down to egg cell.

2. **Pollen grain** is carried to stigma by insects.

Stigma, sticky to catch pollen.

Carpel female part of flower

Style

Egg cell

Anther Pollen is made here.

Stamen Male part of flower

Filament holds the anther up.

Integument

6. **Male and female nuclei** fuse in **fertilisation**.

Ovary (becomes fruit) This contains egg cells which will each become a seed after fertilisation.

5. **Male nucleus** travels down pollen tube.

Apple blossom (flowers)

Sepal

7. **The egg cell** becomes a seed after fertilisation. The ovary becomes a **fruit** which contains seeds.

A petal

Nectary (contains sugar solution)

Questions:
1. Which part of the flower is made up of stigma, style and ovary?
2. What two parts make up the male part of a flower?
3. Where is pollen made?
4. The egg cell contains the female nucleus. Where is the male nucleus?
5. How does the male nucleus reach the female nucleus?
6. Why do insects fly to flowers?
7. Which part becomes a fruit after fertilisation?

METHODS OF POLLINATION Pollination is the transfer of pollen from the anther to the stigma.

Insect-pollinated flower, e.g. buttercup

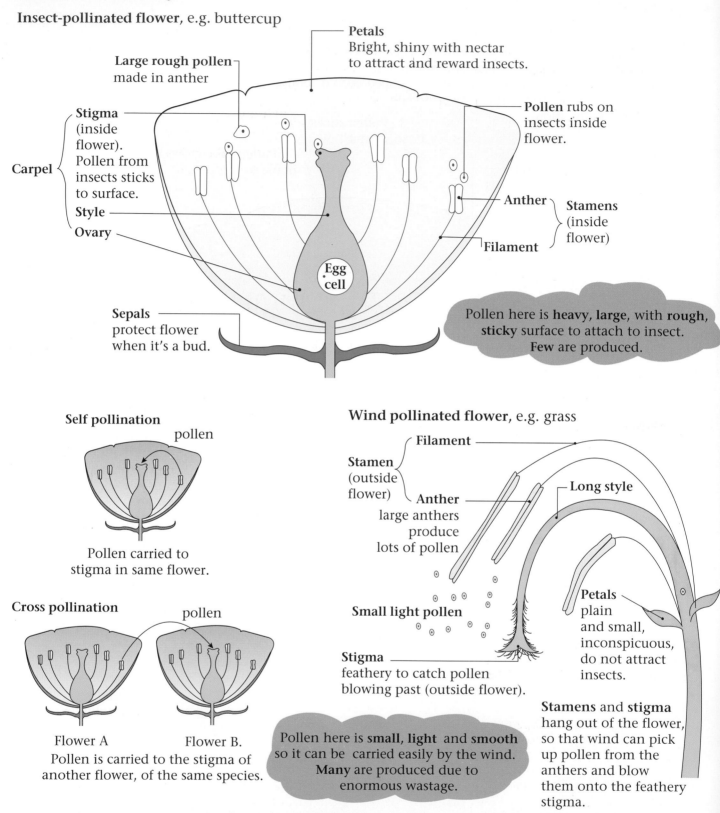

Large rough pollen made in anther

Petals
Bright, shiny with nectar to attract and reward insects.

Stigma (inside flower). Pollen from insects sticks to surface.

Carpel

Style

Ovary

Pollen rubs on insects inside flower.

Anther

Stamens (inside flower)

Filament

Egg cell

Sepals protect flower when it's a bud.

Pollen here is **heavy**, **large**, with **rough**, **sticky** surface to attach to insect. **Few** are produced.

Self pollination

pollen

Pollen carried to stigma in same flower.

Cross pollination

pollen

Flower A.   Flower B.
Pollen is carried to the stigma of another flower, of the same species.

Wind pollinated flower, e.g. grass

Filament

Stamen (outside flower)

Anther large anthers produce lots of pollen

Long style

Small light pollen

Stigma feathery to catch pollen blowing past (outside flower).

Petals plain and small, inconspicuous, do not attract insects.

Stamens and **stigma** hang out of the flower, so that wind can pick up pollen from the anthers and blow them onto the feathery stigma.

Pollen here is **small, light** and **smooth** so it can be carried easily by the wind. **Many** are produced due to enormous wastage.

Questions:
1. What is meant by the term pollination?
2. How does the pollen of a wind pollinated flower differ from an insect pollinated flower?
3. Give 3 other differences between the structure of a wind and insect pollinated flower and explain why these differences are necessary.
4. What is the difference between self and cross pollination and what are the advantages and disadvantages of each method?

GERMINATION This is when a seed starts to grow.

Section through seed

Embryo plant

Testa (seed coat).

Cotyledon (food store for growing embryo).

Cotyledon shrinks as food is used up by growing plant.

Young root (radicle) grows first allowing absorption of *water* and *anchorage.*

Young shoot (plumule) hooked to prevent damage to tip.

Food store almost used up.

Young root

The **cotyledon** contains starch and protein needed by the embryo plant, but starch and protein are insoluble. **Amylase** in the seed digests the starch to sugar and **protease** digests the protein to amino acids. Both sugar and amino acids are soluble and can diffuse to the embryo enabling it to grow first a root and then a shoot. As the food supply is broken down and used, the cotyledon shrinks

Testa shrivels up and disintegrates

Leaves start to make sugar in **photosynthesis.**

Empty seed case; cotyledon all used up.

Root absorbs water and anchors plant.

Conditions needed for germination – an experiment

black paper

A B C D E

seeds

Plant starts to *increase* in *weight* due to *photosynthesis.*

oil
cooled boiled water

wet cotton wool | wet cotton wool in fridge | wet cotton wool | dry cotton wool | wet cotton wool

Tube	Condition	Germination
A	Water, warmth and O_2	Yes
B	No warmth	No
C	Water, warmth and O_2	Yes
D	No water	No
E	No oxygen	No

Seeds need **water, warmth** and **oxygen** to germinate.

Questions:
1. What 3 conditions are needed for seeds to germinate?
2. Explain why the growing seed initially loses weight and then starts gaining weight once it has leaves.
3. Suggest why seeds do not normally need light to germinate.
4. In the experiment, what is the purpose of a) the black paper b) the oil c) the fridge
5. Why didn't the seeds germinate in tube D?

WORDS TO REMEMBER

A

Absorption – The movement of soluble food from the small intestine into the blood.

Acid rain – Acids formed from a combination of air pollutants and water which fall as rain.

Active transport – The movement of molecules from an area of low concentration to an area of higher concentration. This process requires energy.

ADH – Antidiuretic hormone secreted from the pituitary gland which controls the reabsorption of water in the nephron.

Adrenaline – Hormone produced during times of stress to increase the energy available.

Aerobic respiration – Respiration which requires oxygen.

Alcohol – A depressant drug found in drinks.

Algal bloom – The layer of algae covering a lake following eutrophication.

Allele – An alternative form of a gene e.g. T or t.

Alveoli – Air sacs in the lungs where gas exchange takes place.

Anaerobic respiration – Respiration which does not require oxygen.

Anther – where pollen grains are made.

Antibiotics – Chemicals used to kill bacteria.

Antibodies – Chemicals of specific shape produced by lymphocytes which destroy antigens.

Antigens – The surface shape of bacteria which antibodies attack.

Arterioles – Small arteries which link arteries to capillaries.

Artery – Blood vessels carrying blood away from the heart.

Aseptic – sterile conditions required to prevent contamination from other microorganisms in a fermenter.

Asexual reproduction – Reproduction not involving gametes, leading to identical offspring.

Atria – The upper chambers of the heart where blood enters.

Auxins – Plant hormones produced in the shoot and root tip.

B

Benedict's solution – A chemical used to detect reducing sugars such as glucose.

Biological control – Using one species to control the numbers of another.

Biuret solution – A chemical used to detect the presence of protein.

Breathing – The inhaling and exhaling of air, also known as ventilation.

C

Camouflage – The blending of an animal into its habitat.

Capillaries – Tiny permeable blood vessels with walls only one cell thick where exchange of material takes place between the blood and the body cells.

Carbohydrate – A biological molecule including sugar, starch, glycogen and cellulose.

Cardiac muscle – The muscle making up the heart.

Carnivore – Animals which eat other animals.

Carpel – the female part of the plant made up of the stigma, style and ovary.

Cell – The smallest unit of life.

Cell body – The swollen part of the nerve cell containing the nucleus.

Cellulose – Polysaccharide forming the cell wall of plants.

CFC – Chlorofluorocarbons produced by aerosols and fridges causing holes in the ozone layer.

Chloroplasts – Organelle containing chlorophyll where photosynthesis takes place.

Cholesterol – A substance made by the liver and found in the blood. High levels increase the risk of disease of the heart and blood vessels.

Chromosome – A strand of genes, found in the nucleus.

Cilia – Little hairs which beat in a regular way, e.g. in the trachea.

Ciliary muscle – The muscle that can alter the shape of the lens.

Clones – Genetically identical organisms.

CNS – Central nervous system made up of the brain and spinal cord.

Codominance – Two alleles of a pair are equally dominant and both are expressed.

Contraceptive pill – Pill containing oestrogen and/or progesterone to prevent ovulation and therefore pregnancy.

Coronary artery – The blood vessel which provides the heart muscle with oxygen and glucose.

Corpus luteum – Structure in the ovary which develops from the follicle, following ovulation.

Cotyledon – part of a seed which supplies nutrients to the developing plant.

Cuspid valves – Valves between the atria and the ventricles preventing the backflow of blood.

D

Deamination – The breakdown of amino acids to produce urea.

Decay – The breakdown and spoiling of food due to the action of microbes.

WORDS TO REMEMBER

Decomposers – Organisms that feed on dead material, e.g. bacteria and fungi.

Deforestation – The permanent, large scale removal of forests.

Denitrifying bacteria – Bacteria that convert nitrates to nitrogen gas in order to get oxygen.

Depressants – Drugs that slow down nerve transmission and therefore reactions.

Diabetes – A condition caused by lack of insulin and detected by the presence of glucose in the urine.

Diaphragm – The muscle separating the thorax from the abdomen whose movements cause breathing.

Diffusion – The movement of molecules from an area of high concentration to an area of low concentration.

Digestion – The breakdown of large insoluble food to small soluble substances.

DNA – A molecule found in the nucleus which makes up the genes.

Dominant – The stronger allele.

Drugs – Chemicals which alter the way the body works.

Dry mass – The mass of plants once the water is removed by heating to 100°C.

E

Effector – A muscle or gland that brings about an effect.

Egestion – The removal of faeces from the anus.

Embryonic stem cells – Unspecialised cells in the embryo that can be used to replace damaged tissue.

Endangered species – Species whose numbers have declined to low levels.

Endocrine gland – A ductless gland which secretes a hormone into the blood.

Endometrium – The lining of the uterus.

Endotherm – Animals that have a constant body temperature, mammals and birds.

Enzymes – Proteins which speed up chemical reactions, e.g. they speed up the breakdown of food.

Eutrophication – A sudden increase in the nutrient content of lakes or rivers, often as a result of leaching.

Evaporation – The loss of water as water vapour. This leads to cooling.

Evolution – Changes that take place in organisms over a long period of time.

Excretion – The removal of waste made in cells from the body.

Extinct – Species that no longer exist.

Fertilisation – The joining of the ovum and sperm.

Fertility drugs – These help a woman to become pregnant by stimulating the release of ova.

F

Filament – holds up the anther.

Fish farming – The mass production of fish like salmon, in cages.

Flaccid – A plant cell which has lost water and is wilting.

Fluoride – A chemical that hardens enamel and neutralises the acid that causes decay.

Follicle – Structure in the ovary in which the ovum develops.

Food chain – A linear feeding relationship between organisms at each feeding or trophic level, e.g. grass → rabbit → fox.

Food preservation – A method of preventing the decay of food.

Food web – Feeding relationships shown by arrows between many organisms at each trophic level.

Fossils – Remains of animals and plants preserved in rock.

FSH – Follicle-stimulating hormone. This causes the development of the follicle.

G

Gametes – The sex cells e.g. the ovum and sperm.

Gene – A section of DNA that determines our features.

Genetic engineering – Altering genes for a particular purpose e.g. to make insulin.

Genetic fingerprinting – A unique 'bar code' of DNA, used e.g. to identify criminals.

Genotype – The type of genes or alleles present.

Germination – when seeds start to grow.

Glomerular filtrate – The fluid formed by ultrafiltration in the glomerulus.

Glucagon – A hormone secreted by the pancreas which converts glycogen to glucose in the liver.

Glycogen – A store of glucose in the liver.

Greenhouse effect – The warming effect of polluted air.

Growth – The permanent increase in size of an organism.

Guard cells – The specialised cells in leaves which can open and close stomata.

H

Habitat – The place where an organism lives.

Haemoglobin – An iron compound found in red blood cells which combines with oxygen to form oxyhaemoglobin.

Hepatic portal vein – Vein carrying glucose and amino acids from the small intestine to the liver.

Herbivore – Animals which eat plants.

Heterozygote – The 2 alleles of a pair are different e.g. Tt.

Homeostasis – The maintenance of a constant internal environment.

Homozygote – The 2 alleles are the same e.g. tt or TT.

WORDS TO REMEMBER

Hormone – Chemical messenger produced in a gland and carried in the blood to its target organ where it has an effect, e.g. insulin produced in the pancreas has its effect in the liver.

Human Genome Project – A project to discover the complete genetic make-up of humans.

I

Immunity – The ability of an organism to resist an infection.

Implantation – Attachment of the zygote to the uterus wall.

Indicator species – Species whose presence or absence indicates a particular condition, e.g. lichens indicate low levels of SO_2.

Industrial fermenter –large tank in which microorganism grow and reproduce in controlled conditions, producing a desired product.

Insulin – A hormone secreted by the pancreas which converts glucose to glycogen in the liver.

Intercostal muscles – Muscles between the ribs which move the ribs up and down during breathing.

Iodine solution – A chemical used to detect the presence of starch.

Iris muscle – Muscle in the eye which controls the size of the pupil and therefore controls the amount of light entering the eye.

L

Lactic acid – A product of anaerobic respiration in animals.

Leaching – The removal of soluble minerals from the soil by rain.

LH – Luteinising hormone. This causes ovulation.

Limiting factor – A factor in short supply which limits the rate of a reaction.

Lipid – A biological molecule used for insulation, waterproofing and as an energy store.

Lymph – Tissue fluid which drains into lymph vessels.

Lymphocytes – White blood cells which secrete antibodies.

M

Meiosis – A type of cell division that forms the gametes.

Menopause – The stopping of the menstrual cycle in middle-aged women.

Menstrual cycle – The 28 day cycle in females which results in an ovum being produced, controlled by hormones.

Microbes – Microorganisms e.g. bacteria, fungi and viruses.

Microvilli – The folded membrane of a cell that increases the surface area.

Minerals – Substances required by animals and plants for healthy growth.

Mitosis – A type of cell division that leads to growth.

Motor neurone – Nerve cell running from the CNS to an effector.

MRSA – A type of bacteria resistant to many antibiotics.

Mutation – A change in a gene or chromosome.

Myelin sheath – Fatty layer surrounding a nerve cell.

N

Natural selection – The process in which better adapted organisms survive to breed and pass on their useful features, whilst less well-adapted ones do not.

Negative feedback mechanism – A change from the normal level is detected, corrected and returned to the normal level.

Nephron – A tiny tube in the kidneys where the blood is filtered.

Nicotine – A stimulant found in cigarettes.

Nitrifying bacteria – Bacteria that convert ammonia to nitrates in order to gain energy.

Nitrogen-fixing bacteria – Bacteria living in the root nodules of leguminous plants that change nitrogen gas into ammonia.

O

Oestrogen – Ovarian hormone which repairs the uterus lining following the period.

Organ – A group of tissues working together for one function, e.g. the heart is an organ.

Osmoregulation – The control of water level in the body.

Osmosis – The movement of water from an area of high water concentration to an area of lower water concentration through a partially permeable membrane.

Ovaries – Reproductive glands in females producing ova, and the hormones controlling the menstrual cycle, oestrogen and progesterone.

Ovary in plants – part of the carpel containing ovules, the ovary becomes the fruit after fertilisation

Ovulation – The release of the ovum from the ovary.

Ovule – found in the ovary, the ovule contains the female gamete. The ovule becomes the seed after fertilisation.

Oxygen debt – The amount of oxygen needed to break down lactic acid formed during anaerobic respiration.

Ozone – A layer of gas in the upper atmosphere which absorbs harmful UV rays from the sun.

P

Palisade layer – A layer in leaves where most photosynthesis takes place.

Pancreas – A digestive gland secreting enzymes and an endocrine gland secreting the hormones insulin and glucagon.

Pathogens – Organisms that cause disease.

WORDS TO REMEMBER

Penicillin – An antibiotic produced by the Penicillium mould.

Period – The monthly shedding of the uterus wall and some blood.

Peristalsis – The squeezing of food along the gut due to the contraction of muscles.

Pesticides – A chemical used to kill pests, such as insects on crops.

Phagocytes – White blood cells which surround and digest bacteria.

Phenotype – The appearance of the organism.

Phloem – Tubes in the stem which carry sugar and amino acids around a plant.

Photosynthesis – The process in which plants make sugar.

Pituitary gland – Structure in the brain producing the hormones ADH, FSH,LH and growth hormone.

Placenta – Structure which forms in the uterus wall of a pregnant female which attaches the fetus to the mother.

Plasma – The watery part of the blood.

Platelets – Cell fragments that are involved in blood clotting.

Pollen grain – contains the male gamete.

Pollen tube – a tube that grows from the stigma to the ovule transporting the male gamete to the female gamete.

Pollination – the transfer of pollen from the anther to the stigma, either by wind or insects.

Population – A group of the same species together in a habitat.

Predator – Animals that hunt, catch and eat their prey, e.g. hawks.

Predator-prey relationship – The number of predators depends on the number of prey and vice versa. Each controls the size of the others population.

Primary consumer – Animals feeding on the producers, e.g. rabbits.

Producers – Plants producing their own food (sugar) in photosynthesis.

Progesterone – Ovarian hormone maintaining the uterus lining.

Protein – A biological molecule required for growth.

Pupil – The hole in the iris muscle through which light enters the eye.

Pyramid of biomass – A diagram to illustrate the mass of organisms at each trophic level. The size of the box depends on the mass of organisms.

Pyramid of energy flow – A diagram to illustrate the flow of energy between the trophic levels. The size of the box depends on the energy content.

Pyramid of numbers – A diagram to illustrate the number of organisms at each trophic level. The size of the box depends on the number of organisms.

Q

Quadrat – A square frame used to estimate plant populations.

R

Receptor – A sense organ that responds to a stimulus, such as light.

Recessive – The weaker allele.

Red blood cells – Biconcave discs with haemoglobin that carry oxygen.

Respiration – The release of energy from food.

Rings of cartilage – Rings found in the trachea and bronchi to keep the air tubes open.

S

Saprobiotic – Organisms feeding on dead material.

Saturated fats – Hard fats that increase the risk of heart disease.

Secondary consumer – Animals feeding on primary consumers, e.g. foxes.

Selective breeding – A technique used by animal and plant breeders to combine required features and produce a desired variety.

Selective reabsorption – The reabsorption of required materials in the nephron.

Semilunar valves – Valves which close to prevent backflow of blood in veins.

Sensory neurone – Nerve cell running from a receptor to the CNS.

Sewage – Animal waste; faeces and urine.

Sex chromosomes – The pair of chromosomes that determine our sex e.g. XX (female) XY (male).

Sex-linked characteristics – Features caused by genes found on the sex chromosomes.

Sexual reproduction – Reproduction involving gametes joining, leading to variation.

Skeleton – The bony framework of the body.

Solvents – A depressant drug found in glue, lighter fuel and aerosol paints.

Species – Organisms which can interbreed and produce fertile offspring.

Stable population – A population whose numbers remain fairly constant.

Stamen – the male part of the plant made up of the anther and filament.

Starch – A store of carbohydrate found in plants.

Stimulants – Drugs that speed up nerve transmission.

Stigma – part of the carpel which gets sticky, where pollen lands.

Stomata – Holes in the leaf, mainly on the under surface, through which gases can enter and leave a leaf.

T

Temperature regulation – Controlling body temperature.

Tertiary consumer – Animals feeding on secondary consumers.

Testes – Reproductive glands in males producing sperm and testosterone.

WORDS TO REMEMBER

Thermal pollution – The reduction of oxygen in lakes caused by the influx of hot water.

Tissue – A group of similar cells working together e.g. muscle.

Tissue culture – Use of tiny pieces of tissue to artificially grow new plants.

Tissue fluid – The liquid produced by ultrafiltration at the capillary bed which bathes the cells.

Tooth decay – The dissolving of teeth by acid.

Transpiration – The loss of water from a plant, mainly as water vapour through the leaves.

Trophic level – A feeding level in a food chain and in ecology, e.g. producer trophic level.

Tropism – A growth movement in plants in response to a stimulus, e.g. phototropism is a growth movement in response to light.

Turgid – A plant cell full of water and swollen.

U

Ultrafiltration – The process pushing fluid out of the blood caused by high blood pressure.

Urine – Liquid waste excreted from the kidneys.

V

Vasoconstriction – The narrowing of arteries which reduces the volume of blood flowing through it.

Vasodilation – The widening of arteries which increases the volume of blood flowing through it.

Vein – Blood vessels carrying blood to the heart.

Ventricles – The lower chambers of the heart with thick muscle which pump blood out of the heart into arteries.

Venules – Small veins linking capillaries to veins.

Villi – Folds in the surface of the small intestine increasing the surface area for absorption.

W

Warfarin – A chemical that prevents the blood from clotting, used as rat poison.

Water potential – The ability of water molecules to move, a measure of their kinetic energy.

Waxy cuticle – Waxy layer covering a leaf which reduces the loss of water from a leaf.

X

Xerophyte – A plant adapted to dry conditions, e.g. marram grass.

Xylem – Tubes in the stem which transports water and minerals up a plant.

INDEX

A

absorption 3, 82
acid rain 63
active transport 5
adaptations
    animals 50, 51
    plants 50, 125
adrenal glands 96, 97
adrenaline 96, 97
aerobic respiration 94
agriculture 42, 65
air pollution 62, 63
alcohol 95
algal blooms 65
alleles 21, 22, 23, 29
alveoli 8, 9, 90, 91
amino acids 11, 82, 99, 119
Amoeba 3, 7, 9, 94
amylase 81, 129
anaerobic respiration 71, 94
anther 127, 128
anti-diuretic hormone (ADH) 96, 107
antibiotics 69, 95
antibodies 75, 84
antigen 75
aorta 87, 104
appendix 79
aquaculture 66–67
arctic conditions 51
arteries 85, 88, 89
    coronary 86, 87
    pulmonary 86, 87
    renal 104
artificial selection 34
aseptic conditions 32, 33
asexual reproduction 39–41
atria 86, 87
auxins 42

B

bacteria 47, 49, 64
    Coccus 69, 70
    genetic modification 30, 32
    nitrogen cycle 60
    reproduction 70
    sewage treatment 71
    types of 70
    useful and harmful 69
bile 81
bile duct 81
biodiversity 71
biological molecules 11–13
biological pest control 68
biological washing powders 16
biomass pyramid 55
bladder 104
blood
    circulation 9, 85
    clots 74, 83, 86
    exchange of materials 88, 89
    structure 83
    sugar levels 98
    water content 106–107

blood cells 83–84
    red 3, 7, 25, 83, 84, 91
    white 74, 75, 83, 84
blood groups 29
blood pressure 88
blood vessels 85, 88–89
body systems, human 4
body temperature 10, 51, 101–102
brain 76, 112
    blood water control 106–107
    control of breathing 93
    temperature control 102
breathing 92–93
    see also respiration
breeding, selective 34
bronchus 74, 90
buoyancy 13

C

caffeine 95
camels 44
cancer 38
cane toads 53
capillaries 85, 88, 89
carbohydrates 12, 118, 119
carbon cycle 59
carbon dioxide
    in air 59, 61, 62
    exchange in body 88, 91, 92, 93
    and plants 59, 117, 120, 125
carbon monoxide 83, 111
cardiac muscle 86
carnivores 48, 49, 56, 59, 77
carriers 25, 27, 28
cell division 35–38
cells 2–3
    animal 2–3, 7
    blood, see blood cells
    nerve 3, 112, 113
    osmosis 6–8
    plant 2–3, 5, 8, 117, 121–122, 124
    respiration 94
    sex 18, 20, 22, 35–36, 37
central nervous system (CNS) 112
cerebellum 112
cerebral hemisphere 112
cervix 108
chlorofluorocarbons (CFCs) 62, 63
chlorophyll 121
chloroplasts 2, 3, 117, 118, 124
cholesterol 13, 86
chromosomes 18, 19, 20, 22, 35, 37
    crossing over 18, 37
    sex 26, 27, 28
cilia 3, 93
classification of life 47
clones/cloning
    genetic engineering 30
    plants 39–40
    tissue culture 41
clouds 58, 61, 63
Coccus bacterium 69, 70
codominance 29
communities 4
consumers 48, 54

contraceptive pill 110
coronary arteries 86, 87
corpus luteum 110
cotyledon 129
cramp 94
cystic fibrosis 18, 24

D

Darwin, Charles 43
DDT 68
decomposers 59, 60, 64, 65, 70
deforestation 61
dentine 77, 78
desert animals 51
diabetes 16, 96, 98
diaphragm 76, 79, 90, 92
dichlorvos 66
diet 11, 14, 78
diffusion 5, 9, 94, 123
digestion 15, 79–82
digestive system 79–81
diseases
    body defences 74–75
    caused by microbes 69
    inherited 18, 24–25, 27–28
    selection pressure 44
    spread of 73
DNA (deoxyribonucleic acid)
    fingerprinting 31
    mutations 18
    structure and replication 19
    transfer in genetic engineering 30
Dodo 46
dominant alleles 21–23
drugs 95
duodenum 81

E

ecosystems 4
egestion 80
egg (ovum) 3, 18, 26, 35, 108, 110
embryonic stem cells 36
enamel 77, 78
endangered species 61
endocrine system 96–98
    see also hormones
endometrium 109, 110
energy
    losses in food chains 56, 57
    pyramid 56
enzyme-substrate complex 15
enzymes 11, 15–16
    action 15
    commercial uses 16
    digestive 15, 81
    fungi and bacteria 70
ethical problems 31
eutrophication 64, 65, 66
evolution
    evidence for 44, 46
    theory of 43
excretion 103–107
extinctions 46, 61
eye 74, 97, 114–115

INDEX

F

fat, digestion 81
fatty acids 13, 82
feedback, negative 102
fermenter, industrial 32–33
fertilisation 18, 20, 35, 36, 108, 109
fetus 111
fibrin 83
fish 50, 65
fish farming 66–67
Fleming, Sir Alexander 69
flight or fight response 97
flower colour 29
flowers 127–128
fluoride 78
follicle stimulating hormone (FSH)
    96, 110
food
    absorption 3, 82
    carbohydrates 12
    digestion 15, 79–82
    lipids 13
    preservation 72
    proteins 11
food chains 48, 56, 57
food labelling 72
food production, energy losses 57
food tests 14
food webs 48, 49
forests, tropical rain 61
fossil fuels 62, 63
fossil record 46
fossils 44, 45
fructose 12, 16
fungi 47, 49, 64, 69, 70

G

gall bladder 79, 80, 81
gametes (sex cells) 18, 20, 22, 35–36,
    37
gas exchange
    in animals 9, 88, 90–94
    in plants 124–125
gel electrophoresis 31
gene therapy 24
genes 17, 20, 21
genetic counselling 24
genetic crosses 22–23
genetic engineering 30
genetic fingerprinting 31
genetic terms 22
genetic variation 17–18
genotype 23
genus 45
geotropism 42
germination 126, 129
glasshouses 120
global warming 62
glomerulus 105
glucagon 96, 98, 99
glucose 12, 82, 94
    blood levels 98, 99
    tests for 14
glycerol 82

glycogen 12, 97, 98
greenhouse effect 61, 62
growth 35, 38
growth hormone 96
guard cells 117, 124

H

habitats
    pond 50
    woodland/forests 48, 49, 61
haemoglobin 83, 84
haemophilia 18, 27–28
hair colour 21, 22
hairs 100, 101
headlines, biology 1
heart 76, 86–87, 97
heart attack 86
heart valves 87
heat loss 10, 51, 101–102
hepatic artery 81, 85, 99
hepatic portal vein 81, 85, 99
hepatic vein 81, 85, 99
herbivores 49, 56, 57, 59
heterozygote 22, 23
high-density lipoproteins 87
hinge joint 116
homeostasis 96–107
    blood sugar 98, 99
    role of liver 99
    temperature control 101–102
    water in blood 106–107
homozygote 22, 23
hormones 96–98
    adrenal glands 96, 97
    blood sugar control 96, 98, 99
    control of blood water 107
    pituitary 96, 110
    plants 42
    reproduction 96, 110
human body 76
Human Genome Project 19
human growth 38
human life cycle 36
human population, problems caused
    by 71
Huntington's disease 18, 24

I

immunity 75
indicator species 63, 64
ingestion 80
inheritance 17
    codominance 29
    dominant 21, 22, 23, 24
    genetic crosses 22–23
    recessive 21, 23, 24, 25
    sex-linked 27–28
insects, pollination 128
insemination 108
insulin 30, 32, 96, 98, 99
intercostal muscles 90, 92
intestines 3, 76, 79–82
iron 121

J

joints 116

K

kidneys 5, 85, 104–107
kingdoms 47
Kwashiorkor 11

L

lactic acid 94
Lamarck, Jean Baptiste 43
larynx 90
leaching 61, 63
leaves
    adaptations 125
    photosynthesis 117, 118, 125
    water loss 123–124
lens 114, 115
leprosy 111
lichens 63
life cycle
    human 36
    plant (plum tree) 126
light
    and the eye 114, 115
    and photosynthesis 117, 118, 120
lipase 81
lipids 13, 14, 119
lipoproteins 87
liver 76, 79, 81, 97, 99
low-density lipoproteins 87
lungs 9, 74, 76, 86, 87, 90–92, 103
luteinising hormone (LH) 96, 110
lymph vessels 89
lymphocytes 74, 75, 83, 84

M

magnesium 121
malaria 25, 73, 74
malnutrition 11
meat production 57
medulla oblongata 112
meiosis 18, 35, 36, 37
membrane, selectively permeable 6
Mendel, Gregor 23
menstrual cycle 109–110
microbes
    decomposers 59, 60, 64, 65, 70
    fermenter 32–33
    populations 53
    useful and harmful 69
    see also bacteria
micropropagation 40–41
microvilli 3, 82
minerals, plants 5, 121
mitochondria 94
mitosis 35, 37
molecules, biological 11–13
mono-unsaturated fat 13
mosquitoes 50, 73, 74

INDEX

moth, white peppered 44
motor neurones 113
MRSA 69
mucus 93
muscles 94, 97, 116
mutation 18

N

natural selection 43, 44
negative feedback 102
nephrons 104–105
nervous system 112–113
neurones 3, 112, 113
news, biology 1
nicotine 95, 111
nitrates 5, 60, 65, 121
nitrogen 33
nitrogen cycle 60
nitrogen dioxide 63
North Sea fisheries 66
nucleus 2, 3, 19, 20

O

oestrogen 96, 110
oil spills 13
organisation, levels of 4
osmoregulation 105
osmosis 6–8, 89, 122
ovary 36, 96, 108, 110
ovulation 108, 109, 110
ovum (egg) 3, 18, 20, 35, 108, 110
oxygen
    in blood 84, 87
    exchange in body 88, 91–92
    exchange in plants 117, 125
    in water 50, 64, 65
oxygen debt 94
oxyhaemoglobin 83, 84
ozone layer 63

P

palisade cells 3, 117
pancreas 79, 81, 96
Paramecium 53
pathogens 69
    body defences 74, 75, 84
    spread of 73
penicillin 32, 69, 95
peristalsis 80
pest control, biological 68
pesticides 66, 67, 68
petals 127, 128
phagocytes 74, 83, 84
phenotype 22
phloem cells 118
phosphate 121
photosynthesis 12, 59, 61
    in leaf 117, 118, 125
    limiting factors 120
    sugars from 118, 119
phototropism 42
pit-fall traps 52
pituitary hormones 96, 110
placenta 36, 111

plant cells 2–3, 5, 8, 121–122, 124
plants
    adaptations 50, 125
    asexual reproduction 39–41
    carbohydrates 12, 118, 119
    carbon cycle 59
    classification 47
    gas exchange 124–125
    growth 38, 39, 42
    hormones 42
    life cycle 126
    micropropagation 40–41
    minerals 5, 121
    population estimations 52
    proteins 11
    selective breeding 34
    transport systems 5, 118, 122
    water loss 8, 58, 123–125
    see also leaves; photosynthesis
plasma 83
plasmid DNA 30
platelets 83
pleural fluid/membranes 90
polar bear 51
pollen 127, 128
pollination 126, 128
pollution
    air 62, 63
    fish farming 66–67
    pesticides 68
    rivers and lakes 64–65
polydactyly 24
polymerase chain reaction 31
polysaccharides 12
polyunsaturated fat 13
pond habitat 50
pooters 52
populations 4, 52–53
    changes in size 53
    definition 53
    estimating size 52
potassium 121
predator-prey relationship 53
pregnancy 110, 111
primary consumers 48, 54
processed food 72
producers 48, 54, 56
progesterone 96, 110
protease 81, 129
proteins 2, 11
    breakdown 99, 104
    deficiency 11
    sources 11
    tests 14
pulmonary artery 85, 86, 87
pulmonary vein 85, 86, 87
pulp cavity 77, 78
pupil 114
pyloric sphincter 81
pyramids
    biomass 55
    energy 56
    numbers 54

Q

quadrats 52
Queen Victoria 27, 28

R

reabsorption, selective 105, 107
recessive alleles 21, 23, 24, 25
recycling 71
reflex actions 113
renal blood vessels 85, 104, 105
reproduction
    asexual 39–41
    flowering plants 127
    human 36
    sexual 18, 35–36
reproductive systems 74, 96
    female 36, 96, 108–111
    male 96, 108
respiration
    aerobic 94
    anaerobic 71, 94
    plants 125
    see also breathing
respiratory passages 74, 90, 93
retina 114, 115
ribosomes 2
ribs 90, 92
rivers 48, 64–65
root hair cells 3, 5, 121, 122
root nodules 60

S

sabre-toothed cat 77
salivary glands 80
salmon, farmed 66–67
saprobiotic organisms 60, 70
saturated fat 13
sea birds 13
secondary consumers 48, 54
seeds, germination 129
selection
    artificial 34
    natural 43, 44
senses 74
sensory neurones 113
sewage
    pollution 64
    treatment 71
sex cells (gametes) 18, 20, 22, 35–36, 37
sex chromosomes 26, 27, 28
sex-linked inheritance 27–28
shivering 101
sickle cell anaemia 18, 25
skeleton 116
skin 74, 97, 103
    structure 100
    temperature regulation 101, 102
small intestine 81, 82
smoking 93, 95, 111
solvents 95

INDEX

species
    definition and names 45
    endangered 61
sperm 3, 18, 20, 26, 35, 108
spinal cord 113
starch 12, 14, 39, 120
stem cells, embryonic 36
stigma 127, 128
stomach 74, 76, 80, 81
stomata 117, 123–125
sugars 12, 16
    blood levels 98, 99
    from photosynthesis 12, 39, 118, 119
sulphur dioxide 63
surface area, and volume 9–10
survival of the fittest 43, 44
sweat glands 100, 101
sweating 101, 102, 106
synapse 113
synovial joints 116

T

teeth 77–78
temperature
    control of body 10, 51, 101–102
    of Earth 62
    and enzymes 15
    and photosynthesis 120
testes 96, 108
testosterone 96
thalidomide 111
thorax, human 9, 90
thyroid gland 96
thyroxine 96
tissue culture 41
tissue fluid 89
tongue rolling 23
tooth decay 78
tortoiseshell cat 29
trachea 3, 74, 90, 93

transpiration 58, 122, 123
transport, active 5
transport systems 9
    blood 83–85
    plants 5, 118, 122
trees
    life cycle 126
    woodland habitat 49
trophic levels 48
tropical rain forest 61
tropisms 42
Tullgren funnel 52

U

ultrafiltration 105
urea 99, 104
ureter 104
urethra 104
urine 104, 105, 106–107
uterus 36, 108, 109, 110

V

vaccination 75
vacuole 2, 3, 7, 8, 124
vagina 108
variation 17–18
vasoconstriction 88, 101
vasodilation 88, 101
veins 85, 88, 89
    hepatic (liver) 81, 85, 99
    pulmonary 85, 86, 87
    renal 85, 104
vena cava 85, 87, 104
ventricles 86, 87
venules 85, 89
villi 82, 111
viruses 69, 75
vision 115
volume, and surface area 9–10

W

warfarin resistance 44
water
    control of in blood 106–107
    importance to life 58
    life in 50
    in plants 118, 122–125
    pollution of 64–68
    see also osmosis
water cycle 58
weed killers 42
whales 13
wind pollination 128
woodlands 48, 49
    see also forests

X

xerophyte 125
xylem vessels 118, 122

Y

yeast cells
    populations 53
    respiration 94